Index

Other titles from

ISTE

in

Fluid Mechanics

2023

OURABAH Kamel
Collective Phenomena in Plasmas and Elsewhere: Kinetic and Hydrodynamic Approaches

2022

PRUD'HOMME Roger, VINCENT Stéphane
Fluid Mechanics at Interfaces 1: Methods and Diversity
Fluid Mechanics at Interfaces 2: Case Studies and Instabilities

2020

SAD CHEMLOUL Nord-Eddine
Dimensional Analysis and Similarity in Fluid Mechanics

2019

CALTAGIRONE Jean-Paul
Discrete Mechanics: Concepts and Applications

2018

ABABOU Rachid
Capillary Flows in Heterogeneous and Random Porous Media 1

2017

TARDU Sedat
Wall Turbulence Control

2015

FOHR Jean-Paul
Heat and Moisture Transfer between Human Body and Environment

2014

CALTAGIRONE Jean-Paul
Discrete Mechanics

PRUD'HOMME Roger
Flows and Chemical Reactions in Heterogeneous Mixtures
Flows and Chemical Reactions in an Electromagnetic Field

TARDU Sedat
Transport and Coherent Structures in Wall Turbulence

2013

BORGHI Roland, ANSELMET Fabien
Turbulent Multiphase Flows with Heat and Mass Transfer

DÉLERY Jean
Three-dimensional Separated Flow Topology: Critical Points, Separation Lines and Vortical Structures

PRUD'HOMME Roger
Flows and Chemical Reactions in Homogeneous Mixtures

2012

BOUTIER Alain
Laser Velocimetry in Fluid Mechanics
Laser Metrology in Fluid Mechanics: Granulometry, Temperature and Concentration Measurements

PRUD'HOMME Roger
Flows and Chemical Reactions

2011

MORY Mathieu
Fluid Mechanics for Chemical Engineering

TARDU Sedat
Statistical Approach to Wall Turbulence

2010

COLIN Stéphane
Microfluidics

DÉLERY Jean
Handbook of Compressible Aerodynamics

GUINOT Vincent
Waves Propagation in Fluids – 2nd edition

SOULI Mhamed, BENSON J. David
Arbitrary Lagrangian Eulerian and Fluid-structure Interaction

TANGUY Jean-Michel
Physical Processes and Measurement Devices
Mathematical Models
Numerical Methods
Practical Applications in Engineering
Modeling Software

2009

FAVRE-MARINET Michel, TARDU Sedat
Convective Heat Transfer

2008

LALAUZE René
Physical Chemistry of Solid-Gas Interfaces

PEUBE Jean-Louis
Fundamentals of Fluid Mechanics and Transport Phenomena

2007

SCHIESTEL Roland
Modeling and Simulation of Turbulent Flows

Printed and bound by CPI Group (UK) Ltd, Croydon, CR0 4YY

Fluid Mechanics at Interfaces 3

Fluid Mechanics at Interfaces 3

Models and Society

Edited by

Roger Prud'homme
Stéphane Vincent

WILEY

First published 2025 in Great Britain and the United States by ISTE Ltd and John Wiley & Sons, Inc.

ISTE Ltd
27-37 St George's Road
London SW19 4EU
UK

www.iste.co.uk

John Wiley & Sons, Inc.
111 River Street
Hoboken, NJ 07030
USA

www.wiley.com

Library of Congress Control Number: 2025935318

British Library Cataloguing-in-Publication Data
A CIP record for this book is available from the British Library
ISBN 978-1-83669-048-1

Contents

Preface

Interfaces are present in most fluid mechanics problems. They not only denote phase separations and boundary conditions, but also thin flames and discontinuity *waves*.

As with Volume 1, this book is born out of the system and engineering e-journal, *Thermodynamics of Interfaces and Fluid Mechanics,* which studied thin zones separating media with different properties. This includes phase separation as well as thin flames and discontinuity waves. At the macroscopic level, they are compared to material surfaces endowed with thermodynamic properties with their related constitutive laws.

Volume 2 of *Fluid Mechanics at Interfaces*, subtitled *Case Studies and Instabilities*, focused on different kinds of interfaces and flow regimes in their vicinity. We examined cases that involve 1D, 2D or 3D *manifolds* in gaseous and liquid physical states, supercritical fluids and single- or multiphase systems that may be pure or mixed. This volume also consists of six chapters.

Volume 3 of *Fluid Mechanics at Interfaces* focuses on the *Models and Society* for diverse causes. First, we see that models are relative to applications (pollution, drops for propulsion, wind power, microgravity, four-dimensional equations to explain the role of electromagnetic fields). However, the social consequences are important (Lavoisier's chemistry replaced alchemy to explain flames, blood circulation in two loops was a Harvey revolution in medicine, and the fire of weapons led to considerable devastation).

Chapter 1 (Deo Mwelwa, David Ilunga, Jean-Paul Katond Mbay) examines the questions raised by the simulation of a pollutant's concentration degradation in permanent two-dimensional flow using the finite element method. The results obtained show that the size of the impacted area is directly proportional to the speed of propagation and the diffusion coefficient. Chapter 2 (Kwassi Anani, Roger Prud'homme, Mahouton Norbert Hounkonnou) considers an approximate analytical solution for mixed injection regimes that acts on drop vaporization frequency response. Effects due to the variation of the heat transfer coefficient are analyzed for the evaporating mass response factor, calculated on the basis of the Rayleigh criterion. A chaotic response is especially noticed in the process when the heat transfer coefficient is fixed at unity. Chapter 3 (Roger Prud'homme) examines the cases of stationary vortices that can appear inside spherical liquid drops. An incompressible external flow of uniform speed at infinity leads the liquid in the drop by friction to form a Hill vortex. In the second case, the drop is subjected to an axial temperature gradient, causing a variation in surface tension, which can lead to the same Hill vortex. Chapter 4 (Roger Prud'homme) gives a summary of combustion-based weapons and their effects. Flamethrowers and thermobaric weapons, and bombs such as napalm, explosive devices and phosphorus bombs are mentioned. Then, it focuses on missiles, the space army and weapons pollution. Chapter 5 (Roger Prud'homme) is interested in shifting the interface in the spacetime. The macroscopic balance equations of the fluid interfaces were established by considering two scales of length and by means of some approximations. We generalize this method to the case of four interfaces in the presence of electromagnetic fields. Chapter 6 (Roger Prud'homme, Kwassi Anani) discusses the interest that experimentation in microgravity can represent for the study of fluid interfaces. We will limit ourselves to two cases: the first is that of capillary interfaces, where surface tension is present even at equilibrium, and the second is that of thin flames, which only exist outside of equilibrium but which can be considered as generalized interfaces. Chapter 7 (Roger Prud'homme, Mahouton Norbert Hounkonnou, Guillaume Legros) relates how at the end of the 18th century, a discovery linked to combustion could challenge the idea that we had about constituents of matter and lead to radically transforming chemistry. A correct interpretation of this discovery had to be found, and for that, to question existing theories and dogmas. Chapter 8 (Jean-Paul Caltagirone, Roger Prud'homme) is concerned with the modeling of partial wetting by a macroscopic approach in discrete mechanics. The assignment of the surface tension values on the triple line related to the contact angle allows us to consider the partial wetting effects in a dynamic context. Two examples illustrate the validity of this approach. Chapter 9 (Delphin Tomboravo, Roger Vony, Francis Raveloson, Tsialefitry Aly Saandy) states a numerical method of finished differences, making it possible to calculate the variables describing an average flow, no viscous and no heavy, through a mobile wheel of a windmill with a horizontal axis, to deduce its performances from them, by using a grid with irregular steps. The internal stability is ensured by a constraint on the step of temporal discretization in accordance with the CFL condition. Initial conditions are calculated while realizing the flow bypass sections, characteristic quantities initially imposed in the calculation domain volume. Chapter 10 (Roger Prud'homme, François Boustani, Stéphane Vincent) considers circulation in the vessels of the human body. After a brief history, a precise description of the heart, the motor of movement, seemed indispensable, followed by that of the multiple and varied vessels in which the blood flows and exchanges with the organs. This is followed by a fluid mechanics section and an incentive conclusion. Finally, Chapter 11 (Edoh Tossou, Kwassi Anani, Roger Prud'homme) contributes by generalizing the classical series solution for initial boundary value problems of the one-dimensional reaction–diffusion equation on any finite interval of the real line. Finally, the solution of the general problem is obtained as a convergent series over the considered interval, with the construction of an auxiliary. The Hopf–Cole transformation has facilitated the generalization of the exact solution of the Burger's equation to generic intervals, as demonstrated by the method described.

Roger PRUD'HOMME
Stéphane VINCENT
Christian CHAUVEAU
Mahouton Norbert HOUNKONNOU
Kwassi ANANI
April 2025

1

Simulation of a Pollutant's Concentration Degradation in Permanent Two-Dimensional Flow Using the Finite Element Method

In this study, performed a simulation of a pollutant's degradation in two-dimensional flow using the finite element method in an area of 15 by 15 units in length. To this end, we started by establishing the strong formulation of its behavior, and then transforming it to a variational formulation. After defining the data related to the observation domain, we set the initial and final conditions. This allows us to solve the problem. The results obtained show that the size of the impacted area is directly proportional to the speed of propagation and the diffusion coefficient. The pollutant's concentration level only very minimally affects this size.

1.1. Introduction

The economic growth of a country based on mining also involves the proliferation of industries. Certainly, these industries' primary objectives serve humanity by manufacturing diverse substances, thus meeting the needs of society or engineering. They are very often perceived as risky activities resulting, on the one hand, from the constant increase in the power of the production units and, on the other hand, from the combustible, explosive or toxic nature of the products used or manufactured in the factory. This latter toxic nature can be caused by emanations into the environment of gases harmful to human health. Thus, in this

Chapter written by Deo MWELWA, David ILUNGA and Jean-Paul KATOND MBAY.

study, we propose to simulate the degradation of a pollutant in two-dimensional flow by the finite element method. The objectives of these chapters are as follows:

– to determine the impacted areas and the level of pollutant concentration;

– to show the influence of the source output flow on the area impacted by the pollutant.

To achieve this, we will assume that:

– losses by deposition are neglected;

– the pollutant is not radioactive.

1.2. Behavior modeling

As shown in Figure 1.1, the atmospheric flow of pollutants can be modeled by a combined advection and diffusion behavior.

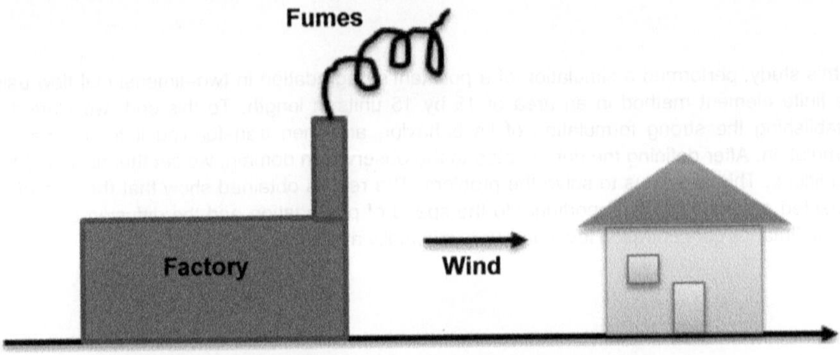

Figure 1.1. *Advection of fumes by wind. For a color version of this figure, see www.iste.co.uk/prudhomme/fluid3.zip*

Mass transport phenomena are easy to observe in our everyday environment. The movement of wind, the spread of a pollutant in the ocean or air, and the slow dispersion of the smell of a perfume are all examples of mass moving from one place to another. Two mass transport phenomena are of particular interest: diffusion and advection (Sportisse 2007).

The diffusion mechanism has two main properties: it is random in nature, and transport is from regions of high concentration to those of low concentration. At the macroscopic level, diffusion often manifests itself as an observable spectrum of intensity or a concentration gradient (see Figure 1.2). Furthermore, the previous points have defined advection as the non-random transport associated with fluid flow. The combination of advection and diffusion is important in several applications of fluid dynamics (Walid 2012).

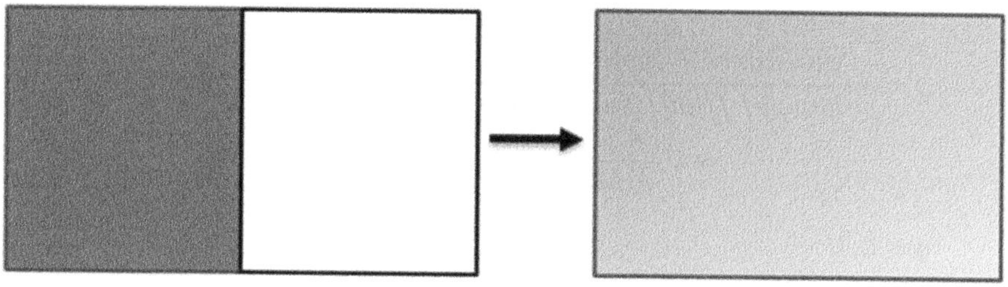

Figure 1.2. *Formation of a concentration gradient. For a color version of this figure, see www.iste.co.uk/prudhomme/fluid3.zip*

Its equation is determined as follows:

The concentration C of a substance in a fluid

The fundamental quantity of interest in environmental fluid mechanics is concentration. In common usage, the term concentration expresses a measure of the amount of a substance in a fluid.

Mathematically, the concentration C is the ratio of the mass M of a substance to the volume of the fluid (Perrin and Scharff 2002):

$$C = \frac{Mass\ of\ the\ substance}{volume\ of\ fluid} = \frac{M}{v}[mg/m^3] \qquad [1.1]$$

The flow

It is not enough to have the concentration in one place; we need to define the amount of substances that move from one place to another. We then define the flow (Perrin and Scharff 2002).

In physics, the flow of a substance in a given direction is defined as the amount passing through a section perpendicular to that direction per unit area and per unit time (Walid 2012).

$$Flow = q = \frac{substance\ passing\ through\ the\ section}{section*time\ duration\ \Delta t} = \frac{M}{A.\Delta t} \qquad [1.2]$$

The advection flow

The advection flow comes from the movement of the carrier fluid. This means that:

$$q = \frac{substance}{volume\ of\ fluid}\frac{volume\ of\ fluid}{surface\ time} = \frac{M}{V}.\frac{V}{A.\Delta t} = \frac{C.V}{V}.\frac{V}{A.\Delta t} = C.\frac{V}{A.\Delta t}$$
$$= C.\frac{A.l}{A.\Delta t}$$

$$q = C.U \qquad [1.3]$$

with U being the driving speed of the carrier fluid.

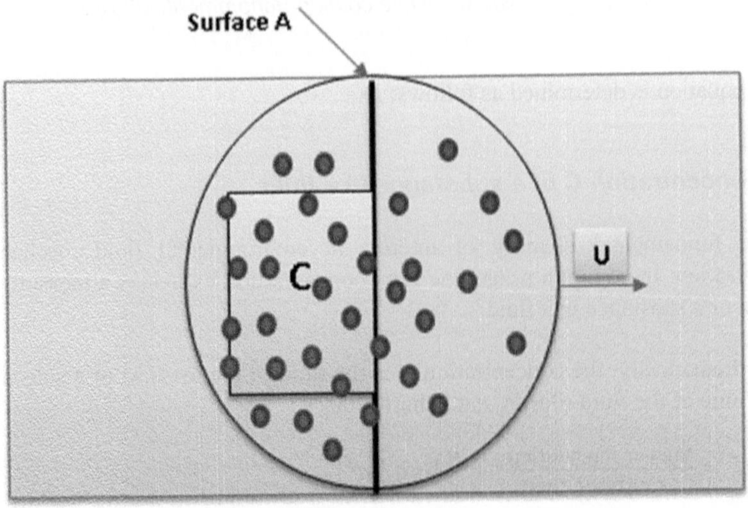

Figure 1.3. *Flow diagram. For a color version of this figure, see www.iste.co.uk/prudhomme/fluid3.zip*

The expression [1.3] means that the advection process takes place by the passive transport of the substance from one place to another with the help of the moving carrier fluid.

The pure advection equation

In pure advection, we are only interested in the advection flow, and we neglect the other flows such as the diffusive flow.

Let us write the flow balance through section A during the time dt in an elementary volume:

$$V = \Delta x. A \qquad [1.4]$$

The variation of mass during a time dt corresponds to the sum of the internal flux (source in V) and the external flux (input and output) (Walid 2012):

$$\begin{bmatrix} the\ variation\ in\ mass \\ in\ V\ during\ dt \end{bmatrix} = \begin{bmatrix} incoming\ flow \\ - \\ outgoing\ flow \end{bmatrix} + \begin{bmatrix} source \\ in\ V \end{bmatrix} \qquad [1.5]$$

$$\begin{bmatrix} \frac{dm}{dt} \end{bmatrix} = \begin{bmatrix} incoming\ flow \\ - \\ outgoing\ flow \end{bmatrix} + \begin{bmatrix} source \\ in\ V \end{bmatrix} \qquad [1.6]$$

Now, the variation in the concentration is given by:

$$dC = \frac{dm}{v} \Rightarrow dm = v.dc \qquad [1.7]$$

A t fixed, the flow entering through section A is q(x,t). A and the outgoing flow through section A is q ((x+dx),t). A, hence the variation of the mass dm in a volume V during a time dt, is as follows:

$$\frac{dm}{dt} = \frac{V.dC}{dt} \qquad [1.8]$$

Replacing [1.4] with [1.6], we get:

$$\frac{A.\Delta x.dC}{dt} = [q(x,t).A] - [q(x + \Delta x, t).A]$$

$$\Rightarrow \frac{dC}{dt} = -\frac{q(x+\Delta x,t)-q(x,t)}{\Delta x} \qquad [1.9]$$

Taking into account the internal flow, i.e. the source r, the balance sheet becomes:

$$\frac{dC}{dt} = -\frac{q(x+\Delta x,t)-q(x,t)}{\Delta x} + r = -\frac{\partial q}{\partial x} + r \qquad [1.10]$$

$$\Rightarrow \frac{\partial C}{\partial t} = -U\frac{\partial C}{\partial x} + r \qquad [1.11]$$

Extrapolating the study to 2D and for a constant velocity U, we obtain the following pure advection equation:

$$\frac{\partial c}{\partial t} + \left(u\frac{\partial c}{\partial x} + v\frac{\partial c}{\partial y}\right) = r \qquad [1.12]$$

Diffusion (Fick's law)

Diffusion is the mechanism of transport of matter under the effect of a concentration gradient, from areas with a concentration of matter to areas with less concentration.

Diffusion is also the process by which a substance is moved from one place to another under the action of random fluctuations (Winiarek 2014).

Consider the two-compartment system (Figure 1.4), and the fluctuating flow compensated by an anti-flow of the same magnitude in the opposite direction.

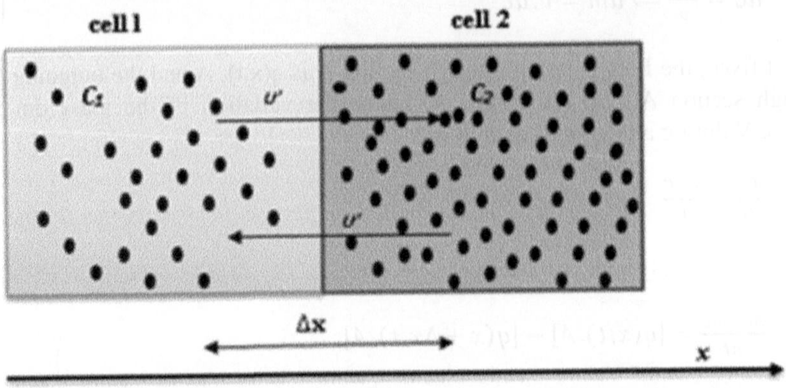

Figure 1.4. *Diagram explaining the broadcast flow. For a color version of this figure, see www.iste.co.uk/prudhomme/fluid3.zip*

Fick's law describes this phenomenon by stating that the flux of matter J is proportional to the concentration gradient via the diffusion coefficient D.

By taking stock of the material flow, we obtain:

$$net\ flow\ of\ 1\ to\ 2 = flow\ of\ 1\ to\ 2 - flow\ of\ 2\ to\ 1 \qquad [1.13]$$

$$J = C_1u' - C_2u' = -u'(C_1 - C_2) = -u'\Delta C = -(u'\Delta x)\frac{\Delta C}{\Delta x} = -D\frac{dC}{dx} \quad [1.14]$$

$$j = -D\left(\frac{dC}{dx} + \frac{dC}{dy}\right) \quad [1.15]$$

where D = diffusion coefficient m^2/s.

In this equation, the minus sign indicates that the transfer takes place from the concentrated areas to the dilute areas.

Pure diffusion equation

In pure diffusion, we are only interested in the diffusive flux, and we neglect the advection flux.

Let us write the flow balance through section A during the time dt in a volume V.

Let j(x,t) be true. A, the incoming flow through A and j ((x+dx),t). A, the outgoing flow.

Starting from equation [1.8], we obtain:

$$\frac{VdC}{dt} = [j(x,t).A] - [j(x + \Delta x, t).A] \quad [1.16]$$

$$\Rightarrow \frac{dC}{dt} = -\frac{j(x+\Delta x,t)-q(x,t)}{\Delta x} \quad [1.17]$$

Taking into account the internal flow, i.e. the source, the balance sheet becomes:

$$\frac{dC}{dt} = -\frac{j(x+\Delta x,t)-q(x,t)}{\Delta x} + r = -\frac{\partial j}{\partial x} + r \quad [1.18]$$

Replacing equation [1.14] in [1.18], we obtain:

$$\frac{\partial C}{\partial t} = -\frac{\partial j}{\partial x} + r = \left(-D\frac{\partial C}{\partial x}\right) + r \quad [1.19]$$

Extrapolating the study to 2D and for a constant diffusion coefficient D, we obtain the following pure diffusion equation:

$$\frac{\partial C}{\partial t} - D\left(\frac{\partial^2 C}{\partial x^2} + \frac{\partial^2 C}{\partial y^2}\right) = r \quad [1.20]$$

Advection–diffusion equation

Adding the advective flow to the diffusive flow, we obtain the following total flux Q:

$$Q = advection\ diffusion\ = C.U - D\ grad\ C\ = Cu\ +\ Cv - D\frac{\partial C}{\partial x} + \frac{\partial C}{\partial y} \quad [1.21]$$

Following the same approach as in the previous cases, we obtain the following advection–diffusion equation:

$$\frac{\partial C}{\partial t} = -\left(\frac{\partial Q}{\partial x} + \frac{\partial Q}{\partial y}\right) + r = -\frac{\partial (Cu)}{\partial x} + \frac{\partial}{\partial x}\left(D\frac{\partial C}{\partial x}\right) - \frac{\partial (Cv)}{\partial y} + \frac{\partial}{\partial y}\left(D\frac{\partial C}{\partial x}\right) + r \quad [1.22]$$

If U, V and D are the constant, we finally have:

$$\frac{\partial C}{\partial t} + u\frac{\partial C}{\partial x} + v\frac{\partial C}{\partial y} - D\left(\frac{\partial^2 C}{\partial x^2} + \frac{\partial^2 C}{\partial y^2}\right) = r \quad [1.23]$$

where C is the concentration of interest in (mg/m^3); u,v is the components of the velocity U along the X,Y axes in (m/s); D is the diffusion coefficient in (m^2/s); r is the source term (mg/m^3.s).

1.3. Evolution of the gas plume concentration

We have demonstrated that the phenomenon of 2D advection diffusion is governed by the partial differential equation [1.23].

Since this equation is nonlinear and does not admit an analytical solution without simplifications, we use one of the approximate methods, called the finite element method, which is developed in the Freefem++ software (Choi 2010).

Solving such an equation using the Freefem++ software is done in the following eight steps (Hecht 2009):

1) definition of the field of study;

2) discretization of the field of study (mesh);

3) definition of the approximation space;

4) definition of problem data and initial conditions;

5) definition of the variational problem on the approximation space;

6) definition of boundary conditions;

7) solving the variational problem;

8) displaying results.

1.3.1. *Definition of the variational problem*

Starting from equation [1.24] defining the strong formulation of the problem to be solved, we apply the finite difference approximation to the term of time derivative, and we obtain:

$$\frac{C_{i,j}^{k+1}-C_{i,j}^{k}}{dt}+u\frac{\partial C}{\partial x}+v\frac{\partial C}{\partial y}-D\left(\frac{\partial^2 C}{\partial x^2}+\frac{\partial^2 C}{\partial y^2}\right)=r \quad [1.24]$$

Or, more simply:

$$\frac{C_{i,j}^{k+1}-C_{i,j}^{k}}{dt}+U.\nabla C_{i,j}^{k}-D.\Delta C_{i,j}^{k}=\mathrm{r} \quad [1.25]$$

with $U=u.\vec{\imath}+v.\vec{\jmath}$

Multiplying equation [1.25] by the test 〖C〗 _h function and applying Green Ostrogasky's theorem, we obtain:

$$Ch.\left[\frac{C_{i,j}^{k+1}-C_{i,j}^{k}}{dt}\right]+Ch.U.\nabla C_{i,j}^{k}+D.\nabla C_{i,j}^{k}.\nabla Ch=r.Ch$$

$$Ch.C_{i,j}^{k+1}-Ch.C_{i,j}^{k}+dt.Ch.U.\nabla C_{i,j}^{k}+dt.D.\nabla C_{i,j}^{k}.\nabla Ch=r.Ch.dt \quad [1.26]$$

Integrating equation [1.26] member by member, we obtain:

$$\iint_{\Omega} Ch.C_{i,j}^{k+1}.dx.dy-\iint_{\Omega} Ch.C_{i,j}^{k}.dx.dy+\iint_{\Omega} dt.Ch.U.\nabla C_{i,j}^{k}.dx.dy + \iint_{\Omega} dt.D.\nabla C_{i,j}^{k}.\nabla Ch.dx.dy=\iint_{\Omega} r.Ch.dt.dx.dy \quad [1.27]$$

The resulting equation [1.27] is the variational (weak) formulation of the mathematical model applied to our system.

1.3.2. *Definitions of initial and boundary conditions*

The notion of boundary conditions is perceived here as the conditions on a boundary delimiting the observation area and not as those of a physical boundary leading to a sudden change of environment. Numerically, we will consider a discrete, closed domain, on which we will perform the concentration calculations. This domain will be bounded by a discrete virtual boundary $\partial\Omega$, one mesh thickness around the Ω domain. The whole therefore constitutes a $\partial\Omega + \Omega$ closed domain. Having in fact no knowledge of the concentration of the $\partial\Omega$ contour, we will set it to zero.

Considering the healthy medium before the emission of the pollutant, we therefore set the initial concentration C_1 in the domain equal to zero.

1.3.3. *Problem data*

The data of the problem are as follows:

– length of the domain along the x-axis: 15 units of length;

– width of the domain along the y-axis: 15 units of length;

– the concentration of the pollutant released from the source is set at 10(mg)/m^3;

– the diffusion coefficient D: 1.24 m^2/s;

– source coordinates: $x = 3.5$ and $y = 3$;

– wind speed: $u = v = 2.7$ m/s;

– presence of some obstacles defined by their boundary conditions.

1.4. Results and interpretation

For a simulation time of 5 units, for a step of 0.5 and for a simulation taking into account the sensitivity of the mesh, the following results are obtained.

We note a deterioration in the concentration of the gas during its propagation (see legend). The result also shows that the areas that are not impacted and therefore out of danger are almost in black. We also see that the propagation of the gas follows the direction of the wind flow. The areas surrounding the source of the emanation are areas of high concentration.

Also, the study of the mesh sensitivity shows that the configuration of the mesh, the result of which is shown in Figure 1.5(c), offers optimal accuracy (result deviation of less than 5%) for an execution time of 9.461 seconds.

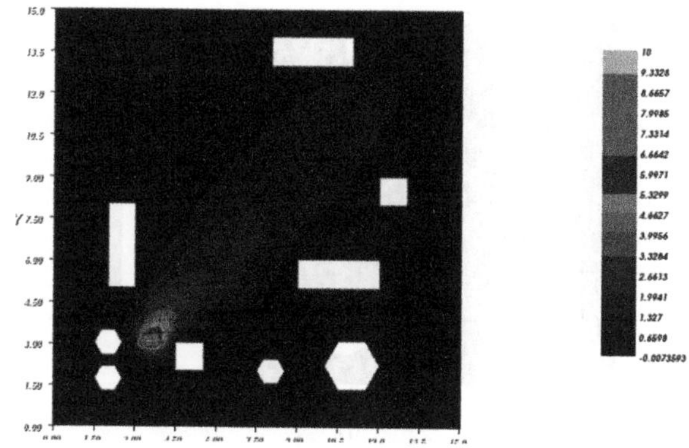
10
9.3328
8.6657
7.9985
7.3314
6.6642
5.9971
5.3299
4.6627
3.9956
3.3284
2.6613
1.9941
1.327
0.6598
-0.0073593

a)

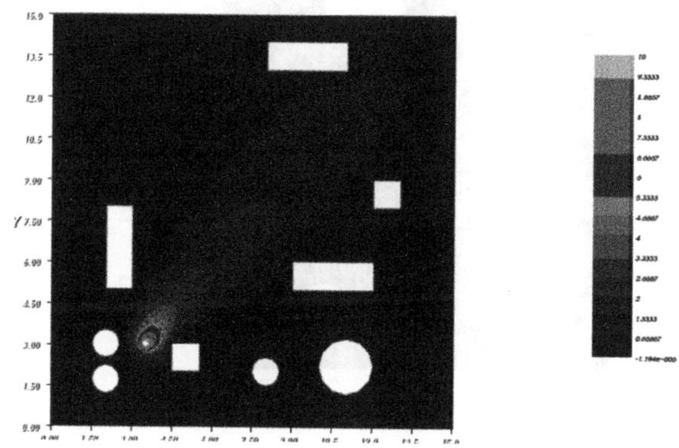

b)

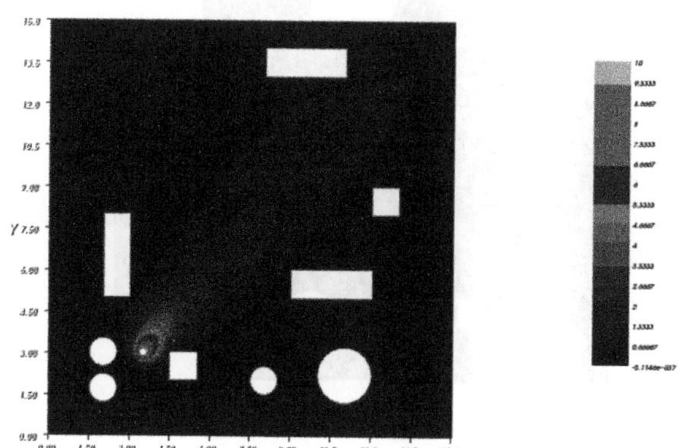

c)

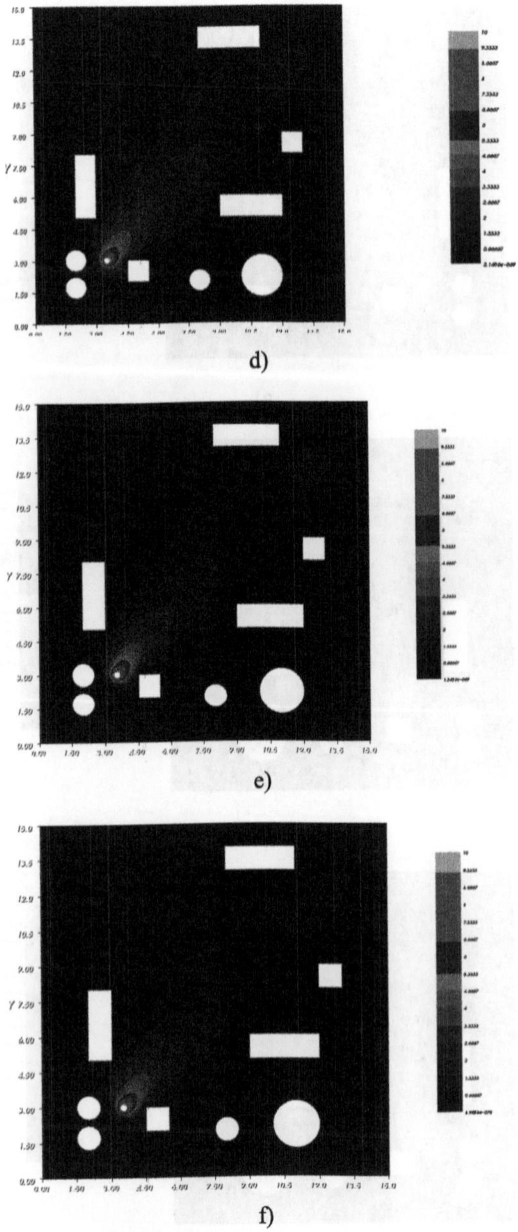

Figure 1.5. *Gas flow simulation for six levels of mesh resolution. For a color version of this figure, see www.iste.co.uk/prudhomme/fluid3.zip*

1.5. Conclusion and perspective

The solution of the advection–diffusion problem shows that the extent of the impacted area is directly proportional to the propagation speed (of the wind) and the diffusion coefficient. The level of concentration of the pollutant affects this extent only very minimally. As a follow-up to this work, we suggest a study of estimating the coordinates of a source in the event of an accidental emanation.

1.6. References

Choi, D. (2010). Méthode des éléments-finis par l'exemple. Document, Université de Caen, Caen.

Hecht, F. (2012). New development in FreeFem++. *Journal of Numerical Mathematics*, 20(3–4), 251–266.

Perrin, R. and Scharff, J.P. (2002). *Chimie industrielle*, 2nd edition. Dunod, Paris.

Sportisse, B. (2007). Modélisation et simulation de la pollution atmosphérique. Thesis, Université Pierre et Marie Curie, Paris.

Walid, M. (2012). Simulation numérique du phénomène d'advection et de diffusion de la concentration d'un polluant dans un écoulement permanent bidimensionnel par la méthode des volumes finis. Master's Thesis, Université Hassiba Benbouali de Chlef, Ouled Fares.

Winiarek, V. (2014). Dispersion atmosphérique et modélisation inverse pour la reconstruction de sources accidentelles de polluants. Thesis, Université Paris-Est, Paris.

2

Drop Vaporization Frequency Response: An Approximate Analytical Solution for Mixed Injection Regimes

The vaporization frequency response due to pressure oscillations is analyzed for a spray of repetitively injected drops into a combustion chamber. In the Heidmann analogy, this vaporizing spray is represented by the so-called "mean droplet", which is a continuously fed spherical droplet at rest inside the combustion chamber. Only radial thermal convection and conduction effects are considered inside the vaporizing mean droplet since the feeding is realized symmetrically with the same liquid fuel, using a point source placed at the center. This feeding process, at some liquid–liquid heat transfer coefficient, is now considered a proper boundary condition of the generalized feeding regime, which controls the whole fuel injection process into the chamber. Effects due to the variation of the heat transfer coefficient are analyzed for the evaporating mass response factor, calculated on the basis of the Rayleigh criterion. A chaotic response is especially noticed in the process when the heat transfer coefficient is fixed at unity.

2.1. Introduction

Combustion instabilities related to pressure oscillations are due to the coupling between acoustic waves and other combustion phenomena such as evaporation, heating, etc. In confined devices, the coupling between the acoustic field and heat or mass release at certain frequency levels may lead to engine failure or other catastrophic consequences (Nair and Sujith 2014). On the contrary, new blends of fuels can be engineered to undergo preferential instabilities leading to homogeneous

Chapter written by Kwassi ANANI, Roger PRUD'HOMME and Mahouton Norbert HOUNKONNOU.

combustion with higher efficiency (Candel et al. 2013). This chapter aims to contribute to the linear analysis of subcritical combustion instabilities in diffusion flame models, via an analytical approach based on the mean spherical droplet configuration (see Anani et al. 2018). In the following section, a brief description is made of the unperturbed state corresponding to the vaporization of the continuously fed spherical droplet in a stable environment. In section 2.3, the linear analysis for harmonic perturbations in pressure is performed, and a double confluent Heun equation (see Slavyanov and Lay 2000) is derived from the energy equation of the liquid phase. An approximate analytical expression of the temperature profile inside the mean droplet is then obtained for the generalized or mixed injection regime, and the mass response factor is defined. The results are discussed in section 2.4 and briefly recalled in the conclusion.

2.2. Stabilized state description

2.2.1. *General assumptions*

Individual spherical fuel droplets are repetitively injected into a subcritical combustion chamber. The distance between the droplets is supposed to be large enough, so that no interaction occurs between the droplets, or between the droplets and the wall. Assuming velocity-stabilized hypotheses as in Heidmann and Wieber (1966), the liquid fuel vaporizing spray is represented in an idealized physical configuration by a mean spherical droplet at rest in the combustion chamber. The mean droplet, placed at a specified location in the combustion chamber (pressure anti-node and velocity node), is supposed to summarize the frequency response of individual drops in the spray. The vaporizing mean droplet has a constant average radius since its instantaneous evaporating mass $\dot{M}$ is continuously restored with an average mass flow rate $\overline{\dot{M}}$ of the same fluid by using a point source placed at the center. The choice of the Arithmetic Mean Diameter configuration is motivated by the analytical approach of the problem since it leads for the mean droplet to conservation equations with fixed boundary conditions. From now on, all barred quantities indicate mean values corresponding to the stabilized state, whereas all primed quantities will denote perturbed quantities, i.e. $x' = (x - \overline{x}) / \overline{x}$.

The local feeding rate $\overline{\dot{M}}$ is distributed throughout the droplet (see Figure 2.1(a)) in such a way that, except for the radial thermal convection effect from the droplet center to its evaporation surface, other convective transport or liquid recirculation phenomenon within the droplet are negligible. The spherical symmetry of the mean droplet is maintained at every moment during the process, and the thermal dilatation of the liquid is negligible so that the density ρ_L, the specific heat c_L and the thermal conductivity k_L of the droplet will be treated as constant. At the mean droplet center, a generalized or mixed boundary condition is considered; that is the liquid fuel is injected with a positive heat transfer coefficient h. The two extreme cases of this injection process are the adiabatic feeding regime ($h = 0$), where zero temperature gradient is assumed at the droplet center, and the isothermal feeding regime ($h = \infty$), where the droplet center is kept at a constant temperature $\overline{T_S}$. The latter is the mean value of the spatially uniform but time-varying temperature T_S of saturated vapor at the stabilized droplet surface.

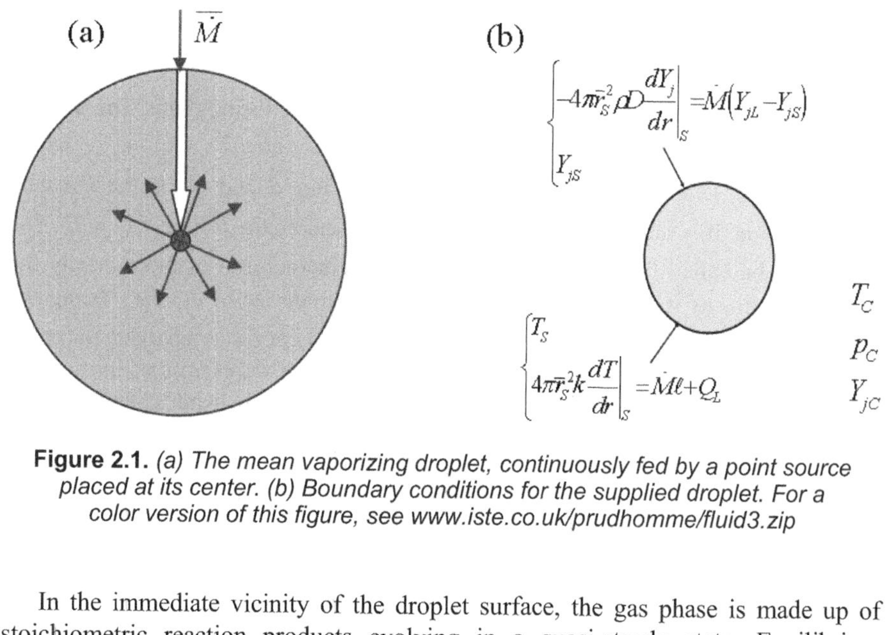

Figure 2.1. *(a) The mean vaporizing droplet, continuously fed by a point source placed at its center. (b) Boundary conditions for the supplied droplet. For a color version of this figure, see www.iste.co.uk/prudhomme/fluid3.zip*

In the immediate vicinity of the droplet surface, the gas phase is made up of stoichiometric reaction products evolving in a quasi-steady state. Equilibrium conditions at the droplet/gas interface are assumed for the stabilized state and there is no gas diffusion into the droplet. Far from the mean evaporating droplet, the ambient environment inside the chamber is at constant subcritical temperature T_C and pressure P_C. The boundary conditions at the supplied droplet surface are shown in Figure 2.1(b). Subscripts L and l refer to liquid phase, whereas subscripts S and C, respectively, indicate the droplet surface and the conditions for the combustion chamber far from the droplet. The heat flux transferred to the liquid is designated by Q_L and the binary diffusion coefficient of fuel vapor in air is denoted by D. The density and the thermal conductivity of the gas mixture around the droplet surface are, respectively, designated by ρ and k. The mass fraction of species j being denoted by Y_j, the gaseous mixture near the surface is composed of fuel species $j = F$ and of combustion products diluted species $j = A$ proceeding from the flame front at infinity. For reason of simplicity, we have considered a mono-component droplet with only fuel species, i.e. $Y_{FL} = 1$ and $Y_{AL} = 0$.

2.2.2. *Characteristic times*

The residence time of the continuously fed droplet can be equated to the mean lifetime of an individual vaporizing droplet in the spray. This time replaces the notion of the free droplet lifetime in the present situation of constant volume and is identified to the ratio $\overline{\tau}_v = \overline{M} / \overline{\dot{M}}$, where $\overline{M}$ represents the mean value of the actual mass M of the supplied droplet and $\overline{\dot{M}}$ is the stationary feeding rate. The transfer time by thermal diffusion process is defined as $\overline{\tau}_T = \overline{r}_S^2 / \kappa_L$, where $\kappa_L = k_L / (\rho_L c_L)$ is the thermal diffusivity of the liquid and $\overline{r}_S$ is the constant average radius. It is then convenient to use the timescale ratio $\theta = 9\overline{\tau}_v / \overline{\tau}_T = \overline{\tau}_v / \tilde{\tau}_T$, which will be called from now on, the thermal exchange ratio or more briefly the exchange ratio, as it is of the same order of magnitude as $1/Pe_L$, Pe_L being the Péclet number of the liquid phase. The coefficient 9 is kept for comparison purposes with results obtained in Anani et al. (2018). During the vaporization, intrinsic or external pressure-related oscillations can cause departure from stabilized-state conditions. The frequency of such ambient pressure oscillations is a major characteristic time of the process. In the case of small harmonic perturbations in pressure, a linear analysis can be performed. The frequency of the harmonic oscillations in ambient pressure will be denoted by ω. In order to provide a parameter depending on the residence time $\overline{\tau}_v$, which may be used to characterize the frequency response for classical fuels, a reduced frequency u defined as $u = 3\omega\overline{\tau}_v$ will be considered.

2.2.3. *Unperturbed state equations*

The mass balance of the mean droplet is:

$$\frac{\mathrm{d}M}{\mathrm{d}t} = \overline{\dot{M}} - \dot{M} \qquad [2.1]$$

with $\overline{\dot{M}}$ denoting the stationary flow of injection and $\dot{M}$ the instantaneous flow of evaporation. In a stabilized state, one has: $\dot{M} \equiv \overline{\dot{M}}$, $\mathrm{d}M / \mathrm{d}t = 0$ and $M = \overline{M}$.

The amount of heat Q_L penetrating into the droplet is expressed as:

$$4\pi\overline{r}_S^2\, k_L \left.\frac{\partial T_l}{\partial r}\right|_{\overline{r}_S,t} = Q_L = Q - \dot{M}\ell \qquad [2.2]$$

where $T_l \equiv T_l(r,t)$ is the temperature value at radial coordinate r and at time t inside the mean droplet. The external gas heat flux is denoted by Q and ℓ designates the latent heat of vaporization per unit mass of the liquid. Equation [2.2] assures the coupling of the gas and the liquid phase solutions at the mean droplet surface. The formulation of the energy conservation equation includes both radial thermal convection and conduction data. In these conditions, the internal temperature T_l satisfies the following equation:

$$\rho_L\, c_L \frac{\partial\, T_l}{\partial t} + \rho_L\, c_L v_r \frac{\partial T_l}{\partial r} - \frac{k_L}{r}\frac{\partial^2 (r\, T_l)}{\partial\, r^2} = 0 \qquad [2.3]$$

where v_r is the central injection velocity expressed as $v_r = \overline{\dot{M}} / 4\pi\rho_L r^2$ with $0 < r < \overline{r}_S$.

This equation is solved, subject to the mixed boundary condition at the droplet center and to the Dirichlet boundary condition at the surface:

$$\begin{cases} \left.\dfrac{\partial T_l}{\partial r}\right|_{r=0,t} = \dfrac{h}{\overline{r}_S}\left(T_l(0,t) - \overline{T}_S\right) \\ T_l(\overline{r}_S, t) = T_S\,(t) \end{cases} \qquad [2.4]$$

The parameter $h > 0$ indicates the heat transfer coefficient. We recall for the adiabatic center injection that $h = 0$ and the mixed boundary condition is reduced to $\partial T_l(0,t)/\partial r = 0$, whereas for the isothermal center injection, $h = \infty$ and the same condition becomes $T_l(0,t) = \overline{T}_S$.

Assuming quasi-steady hypotheses, the droplet surface is in local evaporation equilibrium and the instantaneous mass vaporization rate can be calculated as:

$$\dot{M} = 2\pi\,\rho D r_S\, Sh^* \ln(1 + B_M) = 4\pi \frac{k}{c_p} r_S\, Nu^* \ln(1 + B_T) \qquad [2.5]$$

where $B_M = (Y_{FS} - Y_{FC})/(1-Y_{FS})$ and $B_T = c_p(T_C - T_S)/(\ell + Q_L/\dot{M})$ are the well-known Spalding mass and heat transfer numbers, and c_p the specific heat capacity of fuel vapor at constant pressure. As mentioned above, parameters ρ, k and D are the density, the thermal conductivity and the binary diffusion coefficient of the mixture of vapor and ambient gas. The Sherwood and Nusselt numbers Sh^* and Nu^* were provided by Abramzon and Sirignano (1989) in their extended film model. At the droplet surface, the saturated vapor pressure can be expressed as $p_{sat}(T_S) = \exp(a - b/(T_S - c))$, where a, b and c are some coefficients related to the fuel thermophysical properties. The pressure p_{sat} and the mole fraction X_{FS} of fuel species are connected by the relation $p X_{FS} = p_{sat}(T_S)$, where $p = p_C$ denotes the ambient pressure. If the molecular weight of species j (=A or F) is denoted by M_j, then the mass fraction Y_{FS} of the vapor at the droplet surface can be written as a function of the mole fraction X_{FS} as:

$$Y_{FS} = \frac{\mathrm{M}_F}{\mathrm{M}_F X_{FS} + \mathrm{M}_A X_{AS}} X_{FS} \qquad [2.6]$$

Since concentrations and temperature values are evolving in the gas phase, the averaged properties can be evaluated at some reference concentration $\bar{Y}_j = Y_{jS} + A_r(Y_{jC} - Y_{jS})$ and temperature $\bar{T} = T_S + A_r(T_C - T_S)$ with A_r=1/3. Both Sh^* and Nu^* are assumed equal to two, and the Lewis number $Le = k/\rho D c_p$ is equal to one.

2.3. Linear analyses for small perturbations

2.3.1. *Linear analysis of the liquid-phase equations*

Splitting up the flow variables into steady and unsteady parts can be realized by writing $\Delta f = f - \bar{f}$, where f is a flow parameter, $\bar{f}$ is its mean value, Δf is the absolute perturbation, and $f' = \Delta f/\bar{f}$ is the corresponding relative perturbation. The heat flow at the surface, equation [2.2], is then given by:

$$4\pi \bar{r}_S^2 k_L \bar{T}_S \left.\frac{\partial T_l'}{\partial r}\right|_{\bar{r}_S, t} = Q_L = Q_L - \bar{Q}_L = \Delta Q_L \qquad [2.7]$$

as $\bar{Q}_L = 0$. For the perturbed temperature $T_l^{'}(r,t) = [T_l(r,t) - \bar{T}_l(r,t)] / \bar{T}_l(r,t)$, the energy conservation equation, equation [2.3] can be rewritten as:

$$\frac{\partial (r T_l^{'})}{\partial t} + \kappa_L \left(\frac{3\bar{r}_S}{\theta r} \frac{\partial T_l^{'}}{\partial r} - \frac{\partial^2 (r T_l^{'})}{\partial r^2} \right) = 0 \qquad [2.8]$$

where $\theta = \bar{\tau}_v / \tilde{\tau}_T$ is the thermal exchange ratio (see section 2.2). The perturbed boundary conditions in the mixed feeding regime are deduced from equation [2.4] as follows:

$$\begin{cases} \left. \dfrac{\partial T_l^{'}}{\partial r} \right|_{r=0,t} = \dfrac{h}{\bar{r}_S} T_l^{'}(0,t) \\ T_l^{'}(\bar{r}_S, t) = T_S^{'}(t) \end{cases} \qquad [2.9]$$

Introducing small harmonic perturbations of frequency ω in the form of $f' = \hat{f}(r)\exp(\mathrm{i}\omega t)$, the ambient pressure p_C becomes $p' = \hat{p}_C \exp(\mathrm{i}\omega t)$, while the temperature is expressed as $T_l^{'} = \hat{T}_l(r)\exp(\mathrm{i}\omega t)$, and the heat transferred into the droplet as $\Delta Q_L = \Delta \hat{Q}_L(r)\exp(\mathrm{i}\omega t)$. Equation [2.8] is then transformed into:

$$\mathrm{i} r^2 \omega \hat{T}_l + \frac{3\kappa_L \bar{r}_S}{\theta} \frac{\mathrm{d}\hat{T}_l}{\mathrm{d}r} - \kappa_L r \frac{\mathrm{d}^2 (r\hat{T}_l)}{\mathrm{d}r^2} = 0 \qquad [2.10]$$

or equivalently into:

$$\mathrm{i}\omega \bar{\tau}_T \xi \hat{T}_l + \frac{1}{3\theta\xi} \frac{\mathrm{d}\hat{T}_l}{\mathrm{d}\xi} - \frac{\mathrm{d}^2(\xi \hat{T}_l)}{\mathrm{d}\xi^2} = 0 \qquad [2.11]$$

where $\hat{T}_l$ is taken as a function of the reduced radius variable $\xi = r / \bar{r}_S$ $(0 < \xi < 1)$. The boundary conditions in the generalized feeding regime, equation [2.9], can then be written in connection with ξ as:

$$\begin{cases} \left. \dfrac{\mathrm{d}\hat{T}_l}{\mathrm{d}\xi} \right|_{\xi=0} = \dfrac{h}{\bar{r}_S} \hat{T}_0 \\ \hat{T}_l(1) = \hat{T}_S \end{cases} \qquad [2.12]$$

where $\hat{T}_0$ depends on the initial temperature of the injected liquid fuel.

We now consider the complex number $\overline{s}_0=(1-\mathrm{i})(\omega/2\kappa_L)^{1/2}$, conjugate of $s_0=(1+\mathrm{i})(\omega/2\kappa_L)^{1/2}$, s_0 and $-s_0$ being the roots of the characteristic equation $\mathrm{i}\omega-\kappa_L s^2=0$ obtained from equation [2.10], when neglecting the convective term $(3\kappa_L\overline{r}_S/\theta)\mathrm{d}\hat{T}_l/\mathrm{d}r$. For any given value of the heat transfer coefficient $h>0$, a solution of equation [2.11] subject to conditions equation [2.12] can be sought in the form of $\xi\hat{T}_l(\xi)=J(\xi)\{1-\cos[\overline{s}_0\overline{r}_S\xi\exp(\mathrm{i}\arctan h)]\}$, with $\exp(\mathrm{i}\arctan h)=(\mathrm{i}h+1)/(h^2+1)^{1/2}$, and J referring to a function to be determined. From the second-order truncated expansions of sine and cosine functions that are $\sin(S_0\xi)\approx S_0\xi$ and $\cos(S_0\xi)\approx 1-(S_0\xi)^2/2$ with $S_0=\overline{s}_0\overline{r}_S\exp(\mathrm{i}\arctan h)$, it is deduced that the function ξJ approximately verifies the following double confluent Heun equation:

$$\xi^2\frac{\mathrm{d}^2(\xi J)}{\mathrm{d}\xi^2}+\left(2\xi-\frac{3}{\theta}\right)\frac{\mathrm{d}(\xi J)}{\mathrm{d}\xi}-2\overline{s}_0^2\overline{r}_S^2\frac{h(\mathrm{i}-h)}{h^2+1}\xi^2(\xi J)=0 \qquad [2.13]$$

By using the Maple notation, a solution of equation [2.13] can be expressed as: $J(\xi)=C_0\exp(-3(\theta\xi)^{-1})\,\mathrm{HeunD}(x_1,x_2,x_3,x_4,x)/\xi^{\frac{5}{2}}$, where C_0 is an arbitrary constant and $\mathrm{HeunD}(x_1,x_2,x_3,x_4,x)$ is the double confluent Heun function with its corresponding four parameters: $x_1=0$, $x_2=-[\theta^2(h^2+1)-9-9h^2-24uh(\mathrm{i}h+1)\theta]/4\theta^2(h^2+1)$, $x_3=-[9+(9-24\mathrm{i}u\theta)h^2-24hu\theta]/2\theta^2(h^2+1)$ and $x_4=-[-\theta^2(h^2+1)-9-9h^2-24uh(\mathrm{i}h+1)\theta]/4\theta^2(h^2+1)$ and the variable $x=(\xi^2-1)/(\xi^2+1)$. We recall that the quantity $u=3\omega\overline{\tau}_v$ is the ambient pressure frequency defined in the precedent section. Finally, the condition $\hat{T}_l(1)=\hat{T}_S$ at the mean droplet surface leads for this feeding regime to an approximate analytical solution expressed as:

$$\hat{T}_l(\xi)=\exp\left(\frac{3}{2\theta}\left(1-\frac{1}{\xi}\right)\right)\frac{\hat{T}_S\{1-\cos[\exp(\mathrm{i}\arctan h)\overline{s}_0\overline{r}_S\xi]\}\,\mathrm{HeunD}\left(x_1,x_2,x_3,x_4,\dfrac{\xi^2-1}{\xi^2+1}\right)}{\{1-\cos[\exp(\mathrm{i}\arctan h)\overline{s}_0\overline{r}_S]\}\xi^{\frac{5}{2}}} \qquad [2.14]$$

The above approximate analytical solution presents an essential discontinuity at $\xi = 0$, since, once $h > 0$, the temperature gradient is not null at the droplet center. Now, the calculation of the mass response factor only includes regularity conditions at the droplet surface $\xi = 1$, and these conditions are well verified by this approximate solution.

Thus, the flow condition at the droplet surface (equation [2.7]) can be rewritten as $4\pi\bar{r}_S k_L \bar{T}_S \left.\frac{d\hat{T}_l}{d\xi}\right|_{\xi=1} = \Delta\hat{Q}_L$ and then be applied to the solution equation [2.14]. This leads to:

$$\Delta\hat{Q}_L = -4\pi\bar{r}_S\, k_L \bar{T}_S \hat{T}_S E(u,\theta,h) \qquad [2.15]$$

where E is expressed in the function of u, and h as:

$$E(u,\theta,h) = \bar{s}_0\bar{r}_S \exp(\mathrm{i}\arctan h)\frac{\sin[\exp(\mathrm{i}\arctan h)\bar{s}_0\bar{r}_S]}{\cos[\exp(\mathrm{i}\arctan h)\bar{s}_0\bar{r}_S]-1} - \frac{3}{2\theta} + \frac{5}{2} \qquad [2.16]$$

with $\bar{s}_0\bar{r}_S = (1-\mathrm{i})(3u/2\theta)^{1/2}$, $u = 3\omega\bar{\tau}_v$ and $\theta = \bar{\tau}_v/\tilde{\tau}_T$.

2.3.2. *Gas-phase linearized equations*

The linearized equations for the liquid/gas interface initially presented in Prud'homme et al. (2010) are here briefly recalled. Introducing the harmonic perturbations, the ambient pressure is given by $p' = \hat{p}_C \exp(\mathrm{i}\omega t)$ and the mass flow rate by $\dot{M}' = \hat{\dot{M}} \exp(\mathrm{i}\omega t)$. Consequently, the equations of the gas phase (see section 2.3) imply:

$$\hat{\dot{M}} = \alpha\frac{iu}{1+iu}(\bar{b}\hat{T}_S - \hat{p}_C) \qquad [2.17]$$

and:

$$\Delta\hat{Q}_L = \overline{\dot{M}}\,\bar{\ell}\,(\bar{a}\hat{p}_C - \mu\hat{T}_S) \qquad [2.18]$$

where $u = 3\omega\bar{\tau}_v$ and $\Delta Q_L = \Delta\hat{Q}_L \exp(\mathrm{i}\omega t)$. The coefficients involved in these equations are:

$$\bar{a} = \frac{\bar{T}_C}{\bar{T}_C-\bar{T}_S}\frac{\gamma-1}{\gamma}+\varphi,\ \bar{b}=\frac{\bar{T}_S}{(\bar{T}_S-c)^2}b,\ \mu=\frac{\bar{T}_S}{\bar{T}_C-\bar{T}_S}-\frac{2c}{\bar{T}_S-c}+\bar{b}\varphi \text{ and}$$

$$\alpha=\frac{\bar{B}_M}{(1+\bar{B}_M)\ln(1+\bar{B}_M)}\varphi \text{ with } \varphi=\frac{\bar{Y}_{AC}\bar{Y}_{FS}}{\bar{Y}_{AS}(\bar{Y}_{FS}-\bar{Y}_{FC})}\frac{\mathrm{M}_F}{\mathrm{M}_F\bar{X}_{FS}+\mathrm{M}_A\bar{X}_{AS}}$$

The parameter γ stands for the constant isentropic coefficient and the latent heat of vaporization ℓ per unit mass of the liquid is given by: $\ell = bRT_S^2/\mathrm{M}_F(T_S-c)^2$, where R denotes the universal gas constant.

2.3.3. *Mass response factor*

According to the Rayleigh criterion, when the ambient pressure perturbation $p'=(p-\bar{p})/\bar{p}$ induces a perturbation in the evaporating mass $q'=(q-\bar{q})/\bar{q}$, the mass response factor N is defined as $N=(|\hat{q}|/|\hat{p}|)\cos\phi$, where $|\hat{q}|$ and $|\hat{p}|$ are the moduli of mass release q and pressure p and ϕ is the phase difference between q' and p'. Therefore, a reduced mass response factor can be defined as the real part of the transfer function $Z=\hat{M}/(\alpha\hat{p}_C)$. By using equations [2.15]–[2.18], Z is deduced in function of u, θ and h as:

$$Z(u,\theta,h)=\frac{\mathrm{i}u}{1+\mathrm{i}u}\frac{A+\theta E(u,\theta,h)}{B-\theta E(u,\theta,h)} \qquad [2.19]$$

where $A=3(\bar{a}\bar{b}-\mu)/\lambda$ and $B=3\mu/\lambda$ are some coefficients depending on $\lambda=c_L\bar{T}_S/\bar{\ell}$ and are related to the fuel physical properties. From now on, we will call "response factor" the reduced response factor defined as the real part of the transfer function Z:

$$\frac{N}{\alpha}=\Re(Z) \qquad [2.20]$$

2.4. Results and discussion

In this section, all of the calculations and curves are performed with the fuel thermodynamic coefficients $A=10$ and $B=100$, corresponding approximately to orders of magnitude of values encountered in the classical fuels (Prud'homme et al. 2010). Thus, relative to the heat transfer coefficient h that controls the feeding regime, and to the process characteristic times as defined in section 2.2.2, and again to the influence of the thermodynamic coefficients A and B, the mean droplet mass response factor will be analyzed. Figure 2.2 shows the response factor curves as functions of the reduced frequency $u = 3\omega\overline{\tau}_v$ for arbitrary values of the exchange ratio $\theta = \overline{\tau}_v / \tilde{\tau}_T$. The five columns of diagrams correspond, respectively, to five values of the heat transfer coefficient: $h = 0$; 0.1; 1; 10 and $+\infty$.

First, for $h = 0$ (Figures 2.2(a1), 2.2(a2) and 2.2(a3)) and for $h \to +\infty$ (Figures 2.2(e1), 2.2(e2) and 2.2(e3)), the response factor curves seem, respectively, like those of the adiabatic and of the isothermal injection regimes discussed in Anani et al. (2018). In fact, these curves are identical since, for a given value of the exchange ratio θ, calculations show that:

$$E(u,\theta,h) \to \frac{\overline{s}_0\overline{r}_S\theta\sin(\overline{s}_0\overline{r}_S) + 2\theta\cos(\overline{s}_0\overline{r}_S) - 3\cos(\overline{s}_0\overline{r}_S) - 2\theta + 3}{\theta(1-\cos(\overline{s}_0\overline{r}_S))} = E(u,\theta,0) \qquad [2.21]$$

when $h \to 0$, while

$$E(u,\theta,h) \to -\frac{1}{2}\frac{2\theta s_0\overline{r}_S\sin(s_0\overline{r}_S) + 5\theta\cos(s_0\overline{r}_S) - 3\cos(s_0\overline{r}_S) - 5\theta + 3}{\theta(1-\cos(s_0\overline{r}_S))}$$
$$= E(u,\theta,+\infty) \qquad [2.22]$$

as $h \to +\infty$. The function $E(u,\theta,h)$ is defined by equation [2.16] and mentioned in equation [2.19]. The above expressions of $E(u,\theta,0)$ and $E(u,\theta,+\infty)$ correspond exactly to those used in the calculation of the complex transfer function Z as found for, respectively, adiabatic and isothermal feeding regimes in Anani et al. (2018). Hence, all of the comparative results concerning the two extreme cases of injection highlighted in this latter reference are still valid for the present analysis.

Secondly, according to Figures 2.2(c1), 2.2(c2) and 2.2(c3), the response factor curves show intriguing fluctuations in their profiles when the heat transfer coefficient h is fixed at one. In that case, when the exchange ratio θ is chosen lesser than one, the oscillations become straight chaotic, although they appear relatively reduced in amplitude compared with the cases where θ is much greater than one. Indeed, keeping $h = 1$ and increasing the value of the exchange ratio θ beyond one until a certain threshold value to be hereafter specified, a response factor line exhibits some hyperbolic pattern with high peak value along the reduced frequency axis, as in Figures 2.2(c2) and 2.2(c3). Moreover, once the heat transfer coefficient slightly differs from one, the curves tend to show more lower fluctuations in their profiles, even though h remains very close to one as for $h = 0.95$ or $h = 1.05$, and many other examples not illustrated with figures. For comparison, a unity value of a heat transfer coefficient may characterize a radiation heat transfer processing from the flame to the chamber wall.

According to Santos et al. (2008), for example, the radiative power is highly nonlinear and varies at the first order as the fourth power of the local instantaneous temperature. It may be admitted that, even in fuel injection processes, this specific value of the liquid–liquid heat transfer coefficient ($h = 1$) can strongly influence the evaporating mass release response in a perturbed environment. At this point, experimental investigations are necessary for further clarifications.

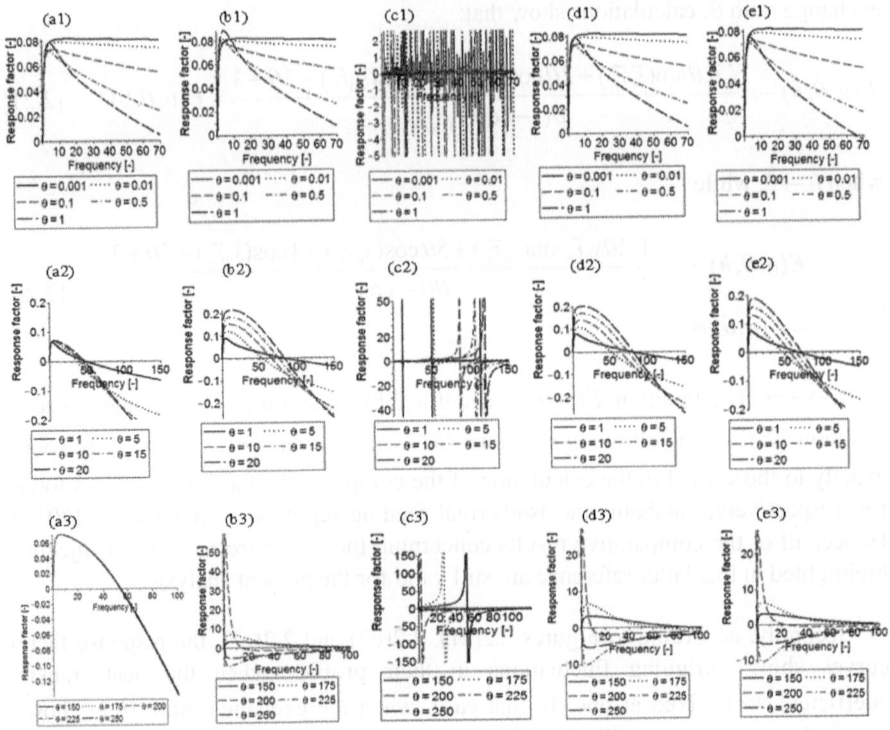

Figure 2.2. *Influence of heat transfer coefficient h on the reduced response factor N/α for different values of the exchange ratio θ for the mean spherical droplet model with A = 10 and B = 100. (a1), (a2) and (a3) for h = 0 or adiabatic center; (b1), (b2) and (b3) for h = 0.1; (c1), (c2) and (c3) for h = 1; (d1), (d2) and (d3) for h = 10; (e1), (e2) and (e3) for h = +∞ or isothermal center. For a color version of this figure, see www.iste.co.uk/prudhomme/fluid3.zip*

Thirdly, for a given value of the exchange ratio θ, almost identical curve profiles are obtained when the not null transfer coefficient h remains much lesser than one as for $0 < h \leq 0.1$. Likewise, when $h \geq 10$, the response curve profiles seem unaffected by the variation of the transfer coefficient h at θ fixed. Indeed, Figures 2.2(d1), 2.2(d2) and 2.2(d3) for $h = 10$ show very similar profiles, respectively, with Figures 2.2(e1), 2.2(e2) and 2.2(e3) for $h = +\infty$. This behavior can be explained by considering the expression of $E(u,\theta,h)$ defined by equation [2.16] and the rate of variation of the term $\exp(\mathrm{i}\arctan h) = (\mathrm{i}h+1)/(h^2+1)^{1/2}$, precisely that of the inner function $\arctan h$ for $h > 0$. Since its rate of variation near $h = 0$ can be equated to one, the function $\arctan h$ tends rapidly enough to zero as h tends to zero. Consequently, the term $\exp(\mathrm{i}\arctan h)$ tends fast to one as h tends to zero. However, when h increases over one, the same function $\arctan h$ becomes asymptotically close to the value $\pi/2$ which is approximately reached as h approaches the value 10. Then, the term $\exp(\mathrm{i}\arctan h)$ tends slowly to the imaginary value i. On the one hand, the first limit leads to the expression of $E(u,\theta, 0)$, giving by equation [2.21] and corresponding to the transfer function Z found in the adiabatic feeding case. As this convergence is rapid, the curve profiles, although unaffected by the variation of h in a deleted neighborhood of zero, seem noticeably different from those obtained for the adiabatic injection regime ($h = 0$). This is readily confirmed by comparisons of Figures 2.2(a2) and 2(a3) for $h = 0$ with Figures 2.2(b2) and 2.2(b3) for $h = 0.1$. On the other hand, the second limit leads to the expression of the function $E(u,\theta,+\infty)$ which intervenes in the isothermal injection transfer function (see equation [2.22]). As the convergence is now asymptotic, the mass response factor curves for $h = 10$ seem very similar to those obtained for the isothermal injection regime ($h = +\infty$).

In brief, we may admit that whenever a not null heat transfer coefficient h is introduced in the injection process, high and nonlinear instabilities may intervene in the vaporization response. As a comparative example, the continuous supply of fuel to the chamber has been theoretically and experimentally identified as an important factor for producing or driving combustion instabilities (Xu et al. 2011; Kannan et al. 2016). Finally, the rapid changes in the curve profiles leading to non-positive response factor representations, especially for $h \geq 10$, are not related to some particular values of the heat transfer coefficient but may rather be related to a specific value of the thermal exchange ratio θ as highlighted in Anani et al. (2018). A more detail analysis also needs to be carried out on this point.

2.5. Conclusion

Through the introduction of a heat transfer coefficient in the liquid fuel injection process, this study has extended, to a more generalized feeding regime, the results on the vaporization frequency response to ambient pressure oscillations. An idealized configuration of a mean droplet has permitted us to compute the mass frequency response for a vaporizing spray. The effects of the liquid heat transfer coefficient are found to be effective for driven or dampen instabilities. Except for the case where the heat transfer coefficient is equal to one, the curves exhibit a single abrupt peak response whenever a positive response exists in the system. The results are also found to be similar to those previously obtained in the adiabatic and isothermal feeding regimes. However, as shown by the present study, a chaotic vaporization frequency response may occur during subcritical combustion processes, provided that certain specific boundary conditions are imposed. The results are also found to be similar to those previously obtained in the adiabatic and isothermal feeding regimes. Indeed, mass response factors in such extreme cases of fuel injection are recovered as simple limit points. The results obtained in this study can be easily generalized to many combustion systems, since the involved parameters are reduced and used in their non-dimensional form. The above-mentioned results may be beneficial for instability control in combustion processes.

2.6. References

Abramzon, B. and Sirignano, W.A. (1989). Droplet vaporization model for spray combustion calculations. *Int. J. Heat Mass Transfer*, 32(9), 1605–1618.

Anani, K., Prud'homme, R., Hounkonnou, M.N. (2018). Dynamic response of a vaporizing spray to pressure oscillations: Approximate analytical solutions. *Combustion and Flame*, 193, 295–305.

Candel, S., Durox, D., Schuller, T., Darabiha, N., Hakim, L., Schmitt, T. (2013). Advances in combustion and propulsion applications. *Eur. J. Mech. B Fluids*, 40, 87–106.

Heidmann, M.F. and Wieber, P.R. (1966). Analysis of frequency response characteristics of propellant vaporization. Technical Report, NASA, Washington, D.C.

Kannan, A., Chellappan, B., Chakravarthy, S. (2016). Flame-acoustic coupling of combustion instability in a non-premixed backward-facing step combustor: The role of acoustic-Reynolds stress. *Combustion Theory and Modelling*, 20(4), 658–682.

Nair, V. and Sujith, R.I. (2014). Multifractality in combustion noise: Predicting an impending instability. *J. Fluid Mech.* 747, 635–655.

Prud'homme, R., Habiballah, M., Matuszewski, L., Mauriot, Y., Nicole, A. (2010). Theoretical analysis of dynamic response of a vaporizing droplet to a acoustic oscillation. *J. Propul. Power*, 26(1), 74–83.

Santos, R.G.D., Lecanu, M., Ducruix, S., Gicquel, O., Iacona, E., Veynante, D. (2008). Coupled large eddy simulations of turbulent combustion and radiative heat transfer. *Combustion and Flame*, 152(3), 387–400.

Slavyanov, S.Y. and Lay, W. (2000). The Heun class of equations. In *Special Functions: A Unified Theory Based on Singularities*. Oxford University Press, New York, 97–162.

Xu, Y., Zhai, M., Dong, P., Wang, F., Zhu, Q. (2011). Modeling of a self-excited pulse combustor and stability analysis. *Combustion Theory and Modelling*, 15(5), 623–643.

3

Considerations about the Hill Vortex

We examine cases of stationary vortices that can appear inside spherical liquid drops. The first case is that of an incompressible external flow of uniform speed at infinity, leading the liquid in the drop by friction to form a Hill vortex. In the second case, the external fluid does not interact by friction with the liquid, however the drop is subjected to an axial temperature gradient, causing a variation in surface tension. This time, it is the induced movement which entrains the internal liquid. Note that the two situations can lead to the same Hill vortex. Combined effects are envisioned. We are also interested in the time factor in these phenomena.

List of symbols

D	Flow rate of a source
$(\vec{\mathbf{e}}_r, \vec{\mathbf{e}}_\theta, \vec{\mathbf{e}}_\varphi)$ $(\vec{\mathbf{i}}, \vec{\mathbf{j}}, \vec{\mathbf{k}})$	Unit basis vectors
K	Intensity of a doublet
$\vec{\mathbf{n}}$	Unit normal
p, p_∞	Pressure, p to infinity
$P_{l,m}$	Legendre polynomial

Chapter written by Roger PRUD'HOMME.

r,θ,φ	Spherical coordinates
R	Gas constant, radius of a sphere
s	Curvilinear abscissa
S	Surface
T	Absolute temperature
$\vec{\mathbf{t}}$	Unit tangent
U, U_∞	Surface velocity, velocity at infinity
$\vec{\mathbf{v}}, v=\lvert\vec{\mathbf{v}}\rvert$	Velocity vector, velocity modulus
$(u,v),(v_r,v_\theta)$	Components of the velocity vector
(x,y,z) (x_1,x_2,x_3)	Cartesian coordinates
$Y_l^m(\theta,\varphi)$	Spherical harmonic
α,α_e,α_i	Stream function coefficients
ϕ	Speed potential
ε	Strain rate
Γ	Intensity of an irrotational vortex
$\mu, \nu=\nu/\rho$	Shear viscosity coefficients
ρ	Mass per unit volume
σ	Surface tension

Ω	Rotation speed
$\vec{\omega}$	Swirl vector
ψ	Stream function
i, e	Resp. internal, external flow
t, n	Resp. tangent, normal to the sphere of radius r
r, θ	Resp. radial, tangential

3.1. Introduction

Spray drops undergo various actions depending on their origin and the resulting physical situation in which they find themselves: watering spray, medical spray, automobile engine injector, rocket engine injector, etc.

Their modeling includes, on the one hand, the examination at the scale of each individual drop and, on the other hand, the study of the spray itself on a larger scale, as a constituent of a multiphase flow, most often liquid–gas.

The individual drop is often considered to be spherical. This is the case for small drops where capillary forces are decisive for establishing sphericity. This simplicity of geometric form is called into question in the case of large drops, as soon as they are subjected to significant forces of aerodynamic origin, for example. *For the sake of simplicity, we try during theoretical research to keep the spherical shape as long as possible.*

The exchanges between the individual drop and its gaseous environment are obviously different depending on whether there is evaporation–condensation or not. They concern the masses of the constituents, the quantity of movement and the energy. Each present phase also undergoes movements and transfer phenomena.

For the flow of the spray itself, we are often led to model what happens inside the individual drops. The most classic is to characterize the latter by their radius, their mass, their temperature and their speed. The distribution in diameter and speed of the drops will always remain the crucial element.

However, it may be interesting to deal with the internal movements of medium-sized drops because these act on the exchanges at their surface. This is the problem that is proposed in this chapter, where we will therefore have to study simultaneously the external and internal flows of individual drops.

We have retained here the case of spherical liquid drops subjected either to a uniform external gas flow or to a thermal gradient in an axial direction. We will see that Hill's vortex modeling is an interesting solution for internal flows.

One of the major problems is the connection between the flow inside the sphere limiting the drop and the flow of the external fluid.

Indeed, if we assume a perfect fluid on the outside, we can logically be led to admit perfect sliding conditions for this fluid at the surface of the drop. But then, how can we admit that there is entrainment of the liquid inside the drop by the external fluid?

The problem of the forces exerted on the drop by the external fluid is also posed with regard to their resultant, which is found to be zero! This constitutes the famous Dalembert paradox. We are then invited to consider the viscosity of the external fluid, at least in the close vicinity of the sphere, which leads to Stokes' theory in the case of the rigid sphere. The results must also be modified to consider a liquid sphere. In the presence of evaporation–condensation of the liquid, it is even more complicated!

Finally, we know that for many linear problems, we can superpose elementary solutions (see Appendix 1 in section 3.6.1). We will do this whenever possible, taking into account the nature of the fluids and the domains of validity that are the interior and exterior of the drop.

3.2. Spherical liquid drop subjected to a uniform external flow at infinity

The Hill vortex is used to model the flow inside a spherical liquid drop in the presence of relative flow (Abramzon and Sirignano 1989).

This vortex is a special case of stationary motion of revolution of a perfect incompressible fluid (Lamb 1945). It is a rotational motion inside a sphere behaving with an external irrotational flow, such that by suitably choosing the multiplicative constant α of the stream function, the speeds of the two flows are identical at the surface of the sphere (Germain 1986).

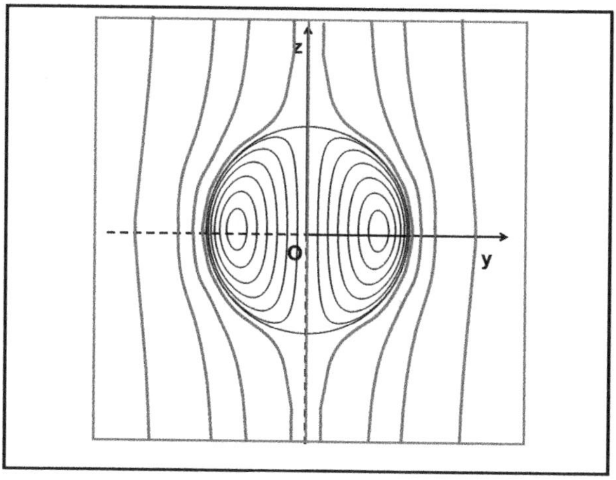

Figure 3.1. *Liquid drop in the presence of a uniform flow at infinity: shape of the streamlines of the external and internal flows in a plane passing through the axis of symmetry. For a color version of this figure, see www.iste.co.uk/prudhomme/fluid3.zip*

We first recall the equations of the fluid flow *outside* the sphere of radius R and then determine the velocity field of the compatible[1] stationary liquid flow *inside* this same sphere (Prud'homme 2012).

The steady flow of a perfect fluid inside a sphere of radius R and the flow around this sphere are characteristic examples, shown in Figure 3.1. A classical presentation of the external flow is given in the appendices (section 3.6).

3.2.1. *Incompressible fluids in spherical coordinates*

3.2.1.1. *Presentation of the coordinate system*

The quantities r, θ,φ are the spherical coordinates represented in Figure 3.2.

In the case of symmetry around the Oz axis, we work in the plane at $\phi = Const$ of Figure 3.2 since the motion is independent of this angle. We can define the basic unit vectors there.

1 That is, without any discontinuity in speed at the interface.

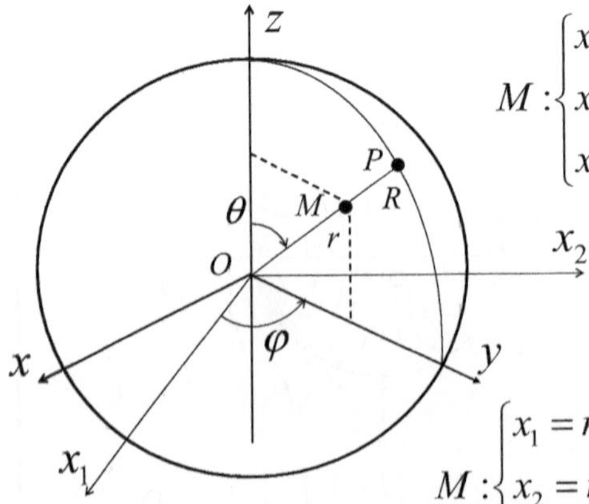

$$M:\begin{cases} x_1 = y\cos\varphi \\ x_2 = y\sin\varphi \\ x_3 = z \end{cases}$$

$$M:\begin{cases} x_1 = r\sin\theta\cos\varphi \\ x_2 = r\sin\theta\sin\varphi \\ z = r\cos\theta \end{cases}$$

Figure 3.2. *Definition of spherical coordinates and representation of a sphere of radius R. The current point M can be inside or outside the sphere considered. P is the point of the OM half-line located on the surface of the sphere*

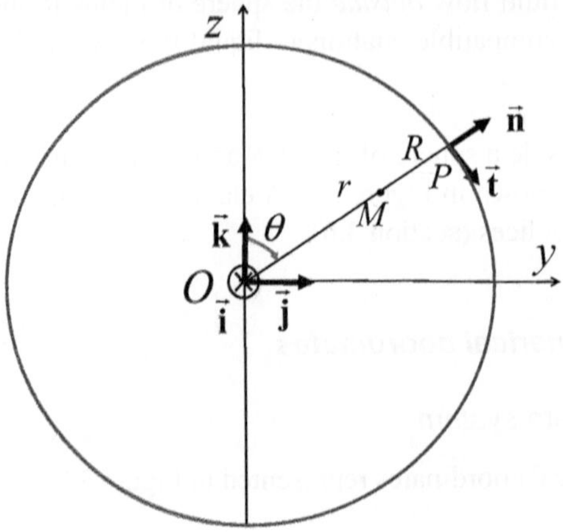

$$M\begin{cases} y = r\sin\theta \\ z = r\cos\theta \end{cases}$$

$$P\begin{cases} y = R\sin\theta \\ z = R\cos\theta \end{cases}$$

$$\begin{cases} \vec{\mathbf{i}} = \vec{\mathbf{j}}\times\vec{\mathbf{k}} \\ \vec{\mathbf{j}} = \vec{\mathbf{t}}\cos\theta + \vec{\mathbf{n}}\sin\theta \\ \vec{\mathbf{k}} = -\vec{\mathbf{t}}\sin\theta + \vec{\mathbf{n}}\cos\theta \end{cases}$$

Figure 3.3. *Definition of coordinates in the y0z plane containing the axis of symmetry Oz. At point P of the surface of the sphere of radius R, we define the unit vectors tangent t and normal* $\vec{n}$ *to the great circle considered. These vectors correspond to the base vectors* $(\vec{e}_\theta, \vec{e}_r)$ *at the point considered*

3.2.1.2. *Continuity equation*

The continuity equation of an incompressible fluid is written in vector notation: $\vec{\nabla} \cdot \overline{\mathbf{v}} = 0$. In expanded form, this gives:

$$\frac{1}{r^2}\frac{\partial}{\partial r}\left(r^2 v_r\right)+\frac{1}{r\sin\theta}\frac{\partial}{\partial\theta}\left(\sin\theta\, v_\theta\right)+\frac{1}{r\sin\theta}\frac{\partial v_\varphi}{\partial\varphi}=0 \qquad [3.1]$$

The values of the various quantities of the fluid no longer depend on the angle φ. The continuity equation is therefore reduced to: $\frac{1}{r^2}\frac{\partial}{\partial r}\left(r^2 v_r\right)+\frac{1}{r\sin\theta}\frac{\partial}{\partial\theta}\left(\sin\theta\, v_\theta\right)=0$, and we can then define the velocity field from the current function ψ as follows:

$$v_r = -\frac{1}{r^2\sin\theta}\frac{\partial\psi}{\partial\theta}, \quad v_\theta = \frac{1}{r\sin\theta}\frac{\partial\psi}{\partial r}, \quad v_\varphi = 0 \qquad [3.2]$$

3.2.1.3. *Expressions of the current function in the presence of a sphere*

We consider two kinds of flows: an irrotational flow (e) outside the sphere of radius R, and a rotational flow (i) of vorticity ω inside the sphere. The flows connect at any point of their spherical boundary.

Regarding vorticity, zero outside, we demonstrate that $\omega = 5\alpha_i y$ by virtue of the transport law of the vorticity vector $\vec{\boldsymbol{\omega}}$ of the internal fluid.

In the spherical coordinate system described above, we have: $r^2 = y^2 + z^2$.

The stream functions ψ_e and ψ_i around and inside the sphere of radius R are expressed as follows[2]:

$$\begin{cases}\psi_e = \alpha_e y^2\left(1-\frac{R^3}{r^3}\right), \text{ with } \alpha_e = V_0/2 \text{ for } r \geq R, \text{ and} \\ \psi_i = \alpha_i y^2\left(R^2 - r^2\right), \text{ with } \alpha_i = -3V_0/4R^2 \text{ for } r \leq R\end{cases} \qquad [3.3]$$

then:

$$\begin{cases}\psi_e = \frac{V_0}{2} y^2\left(1-\frac{R^3}{r^3}\right) & \text{for r} \geq R, \\ \psi_i = -\frac{3V_0}{4} y^2\left(1-\frac{r^2}{R^2}\right) & \text{for } r \leq R\end{cases}. \qquad [3.4]$$

2 Germain (1986) p. 312.

In the problem of motion around a stationary sphere of radius R, the velocity V_0 is the velocity at infinity of the external flow: $V_0 = U_\infty$.

The calculations performed with the current function are summarized in Table 3.1.

	Internal flow (Hill vortex) $r \le R$	External flow $r \ge R$
Stream function	$\psi_i = \alpha_i r^2 \sin^2\theta\left(R^2 - r^2\right)$	$\psi_e = \alpha_e r^2 \sin^2\theta\left(1 - R^3/r^3\right)$
Velocity $\vec{\mathbf{V}}$	$\begin{cases} v_{ri} = -2\alpha_i\left(R^2 - r^2\right)\cos\theta \\ v_{\theta i} = 2\alpha_i\left(R^2 - 2r^2\right)\sin\theta \end{cases}$	$\begin{cases} v_{re} = -2\alpha_e\left(1 - R^3/r^3\right)\cos\theta \\ v_{\theta e} = 2\alpha_e\left(1 + R^3/2r^3\right)\sin\theta \end{cases}$
Coefficients	$\alpha_i = U_S/2R^2 = -3V_0/4R^2$	$\alpha_e = -U_S/3 = V_0/2$

Table 3.1. *Correspondence between the coefficients for an external flow coming from negative x with the velocity module U_∞. U_S is the velocity at the surface of the sphere at z = 0. This table integrates the results from section 3.2*

3.2.1.4. *Flow outside the sphere*

The stationary flow of perfect fluid outside the sphere is an irrotational flow. Its current function (Figure 3.3) is (index *e* for outside):

$$\psi_e = \alpha_e r^2 \sin^2\theta\left(1 - \frac{R^3}{r^3}\right), \quad r \ge R, \quad \alpha_e = \frac{U_\infty}{2} \qquad [3.5]$$

The components of the speed vector are:

$$v_{re} = -2\alpha_e\left(1 - R^3/r^3\right)\cos\theta, \quad v_{\theta e} = 2\alpha_e\left(1 + R^3/2r^3\right)\sin\theta \qquad [3.6]$$

Its streamlines are shown in Figure 3.3 with:

$$\bar{z} = z/R, \bar{y} = y/R, \bar{\psi}_e = 2\psi_e/U_\infty R^2. \qquad [3.7]$$

Maximum speed: let us call U_S the value of the speed at $z=0,\ y=R$, which corresponds to $r=R, \theta=\pi/2$. We find: $U_{Si}=U_{zi}=2\alpha_i R^2, U_{yi}=0$. If we want to be positive, we must therefore have to be positive. We then have $\alpha_i = U_S/2R^2$: the current function is written as $\psi_i = \frac{U_S}{2R^2} r^2 \sin^2\theta\left(R^2 - r^2\right)$. U_S is the maximum value of U at the surface of the sphere.

The velocity vectors of this flow and of the internal flow will be identical at all points on the surface of the sphere if the velocity at infinity is suitably chosen $U_{re}(\infty) = -U_{re}(-\infty) = U_\infty$.

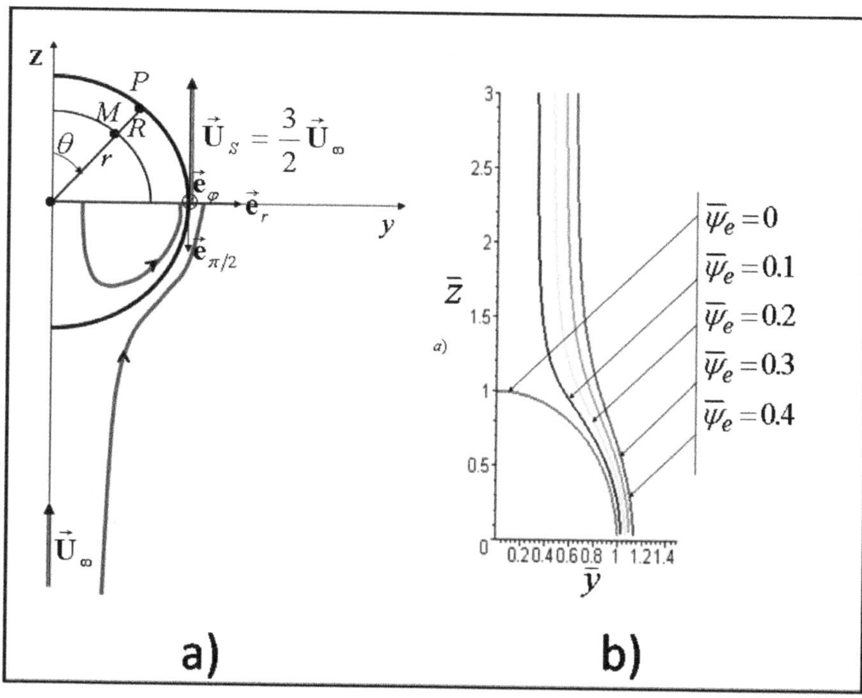

Figure 3.4. *a) Inner and outer streamlines, velocity vectors at infinity $\vec{U}_\infty$ and maximum U_S at the surface. b) Streamlines of the flow outside the sphere in the upper quarter plane passing through the axis. $\bar{z} = \bar{z}(\bar{y})$, for different values of the reduced stream function $\bar{\Psi}_e = 2\Psi_e / U_\infty R^2$. For a color version of this figure, see www.iste.co.uk/prudhomme/fluid3.zip*

3.2.2. *Flow inside the sphere*

3.2.2.1. *Hill vortex*

The stream function of the form:

$$r \le R, \quad \psi_i = \alpha_i r^2 \sin^2\theta\left(R^2 - r^2\right) \qquad [3.8]$$

corresponds to an internal rotational flow (Germain 1986). Elements on rotational flows of perfect rotational fluids are given in Appendix 2 in section 3.6.2.

Knowing the stream function of the internal flow, we find the following radial and angular components of the velocity vector:

$$v_{ri} = -2\alpha_i\left(R^2 - r^2\right)\cos\theta, \quad v_{\theta i} = 2\alpha_i\left(R^2 - r^2\right)\sin\theta \qquad [3.9]$$

The results are shown in Figure 3.4.

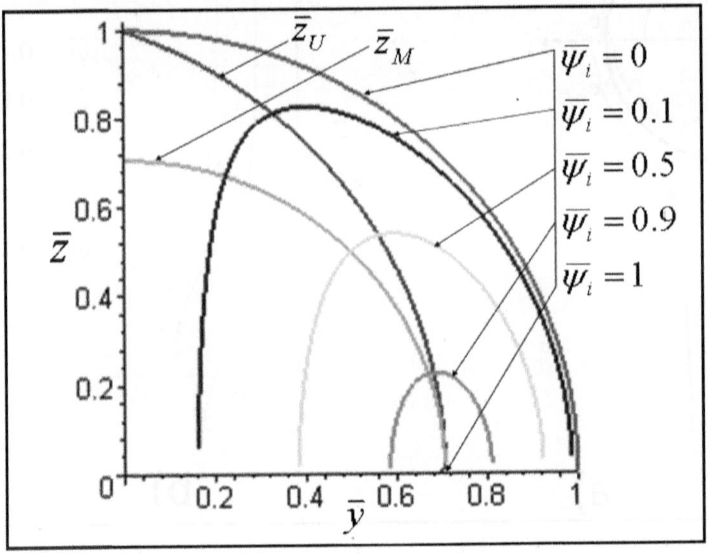

Figure 3.5. *Streamlines calculated in the upper quarter plane passing through the axis* $\overline{z} = \overline{z}(\overline{y})$. $\overline{y} = y/R, \overline{z} = z/R$, *for several values of* $\overline{\Psi}_i = 4\Psi_i / \alpha R^4$. $\overline{z}_m(\overline{y})$ *corresponds to the geometric locus of the points of contact of the lines from O tangent to the streamlines. It is a quarter circle of radius* $R\sqrt{2}/2$. $\overline{z}_u(\overline{y})$ *is the geometric locus of the points where the speed is minimum. For a color version of this figure, see www.iste.co.uk/prudhomme/fluid3.zip*

Each value of the stream function corresponds to the toric surface. The limiting value $\overline{\psi}_i = 0$ is the sphere itself. The velocity vectors of the external flow are identical to those of the Hill vortex at any point on the surface of the sphere if the constants α_i and α_e are chosen appropriately and in relation to the velocity at infinity U_∞ (we have: $v_{re}(\infty) = -v_{re}(-\infty) = U_\infty$).

In this case, we find:

$$\alpha_i = 3U_\infty / 4R^2 \, , \; \alpha_e = -U_\infty / 2 \qquad [3.10]$$

Let us call U_S the value of the *maximum speed at the surface of the sphere*, which occurs at $y = R, z = 0$, which corresponds to: $\theta = \pi/2$. We find: $U_{Si} = 2\alpha_i R^2$. If we want U_S to be positive, we must therefore have α_i to be positive. We then have: $\alpha_i = U_S / 2R^2$, and the current function is written as:

$$\psi_i = \frac{U_S}{2} r^2 \sin^2\theta \left(R^2 - r^2 \right) \qquad [3.11]$$

We can therefore determine the velocity field and the current surfaces of the internal and external flows of the sphere (Germain 1986). We have shown that the flows are then compatible if $\alpha_i = 3U_\infty / 4R^2$. The maximum velocity at the surface of the sphere could then be determined. The details of the calculations in the spherical drop are given in Appendix 3 in section 3.6.3.

3.2.2.2. *Vorticity in viscous fluid*

The balance equation of the vorticity vector in incompressible viscous fluid is written as:

$$\partial \vec{\boldsymbol{\omega}} / \partial t + \vec{\nabla} \times \left(\vec{\nabla} \times \vec{\boldsymbol{\omega}} \right) = \nu \Delta \vec{\boldsymbol{\omega}}$$

with, in spherical coordinates:

$$\omega = \frac{1}{2r} \left[\frac{\partial (r v_\theta)}{\partial r} - \frac{\partial v_r}{\partial \theta} \right] \vec{\mathbf{e}}_\varphi$$

Internal flow: we find it without difficulty, for the Hill vortex: $\vec{\boldsymbol{\omega}}_i = -5\alpha_i r \sin\theta \, \vec{\mathbf{e}}_\phi$.

The intensity ω_t of the vorticity vector is proportional to the distance $y = r\sin\theta$ from the axis of symmetry. The second term of the vorticity vector balance is calculated as follows. We set:

$\vec{\mathbf{A}} = \vec{\nabla}\times\vec{\boldsymbol{\omega}}$ and we obtain:

$$A_r = 10\alpha_i r\sin^2\theta\left(R^2 - 2r^2\right),\quad A_\theta = 10\alpha_i r\sin\theta\cos\theta\left(R^2 - r^2\right).$$

We then find: $\vec{\nabla}\times\vec{\mathbf{A}} = \vec{\nabla}\times\left(\vec{\nabla}\times\vec{\boldsymbol{\omega}}\right) = \dfrac{1}{2r}\left[\dfrac{\partial(r A_\theta)}{\partial r} - \dfrac{\partial A_r}{\partial\theta}\right]\vec{\mathbf{e}}_\phi = \vec{\mathbf{0}}$.

In steady flow: $\dfrac{\partial\vec{\boldsymbol{\omega}}}{\partial t} = \vec{\mathbf{0}}$. It follows that the term of the second member $\nu\,\Delta\vec{\boldsymbol{\omega}}$ is also zero.

The consequence is that the Hill vortex, a perfect fluid solution, is also a solution of the viscous fluid equations.

External flow: we obviously have $\vec{\boldsymbol{\omega}} = \vec{\mathbf{0}}$ since the flow is irrotational.

3.3. Flows of a spherical liquid drop subjected to an axial thermal gradient

3.3.1. *Presentation of the problem*

The effect of an imposed axial temperature field on the internal movements of a liquid spherical drop has been studied by various authors. These movements are caused by the induced surface tension variations and by the Marangoni effect which results in a viscous fluid. Bauer can be cited in particular (Bauer 1982; Bauer 1985; Bauer and Eidel 1987). For the mathematical analysis, spherical harmonics are generally used (see Appendix 4 in section 3.6.4).

In the paper by Bauer (1982), a freely floating liquid drop is subjected at its surface to an axial temperature field inducing Marangoni thermal convection due to the variation of the surface tension. The stream function and the velocity distribution are determined analytically for the stationary and nonstationary temperature fields, by solving the equation satisfied by the stream function using the associated Legendre functions of the first type. The special case of a stable linear axial temperature field is evaluated numerically. We consider the case of axial symmetry Oz, the liquid drop being centered at O. The equation of continuity of the flow of the liquid assumed to be incompressible is written as:

$$\frac{1}{r^2}\frac{\partial}{\partial r}\left(r^2 u\right)+\frac{1}{r\sin\theta}\frac{\partial}{\partial\theta}(v\sin\theta)=0 \qquad [3.12]$$

where u is the radial velocity and v the angular velocity. We can therefore introduce the current function ψ such that:

$$u=-\frac{1}{r^2\sin\theta}\frac{\partial\psi}{\partial\theta},\ v=\frac{1}{r\sin\theta}\frac{\partial\psi}{\partial r} \qquad [3.13]$$

By eliminating the pressure of the unsteady equations of momentum, and defining the operator:

$$\overline{\Delta}=\frac{\partial^2}{\partial t^2}+\frac{\sin\theta}{r^2}\frac{\partial}{\partial\theta}\left(\frac{1}{\sin\theta}\frac{\partial}{\partial\theta}\right) \qquad [3.14]$$

it becomes:

$$\overline{\Delta}\left[\overline{\Delta}\psi-\frac{1}{\nu}\frac{\partial\overline{\psi}}{\partial t}\right]=0 \qquad [3.15]$$

Bauer first treats the stationary case by assuming the derivative of the surface tension to be constant σ relative to the temperature T.

This temperature develops into a series of Legendre polynomials as follows:

$$T_0+T_1\ f_0(\xi)=\sum_{n=0}^{\infty}\alpha_n\ P_n(\xi) \qquad [3.16]$$

with $\xi=\cos\theta$ and the orthogonality relation:

$$\int_{\xi=-1}^{+1}P_m(\xi)P_n(\xi)d\xi=\begin{cases}0 & \text{for}\quad m\neq n\\ \dfrac{2}{2n+1} & \text{for}\quad m=n\end{cases} \qquad [3.17]$$

The temperature distribution inside the spherical drop is given by:

$$T(r,\theta)=\sum_{n=0}^{\infty}\alpha_n\left(\frac{r}{R}\right)^n P_n(\cos\theta) \qquad [3.18]$$

with the coefficients α_n verifying:

$$\alpha_n = \left[T_0 \int_{-1}^{+1} P_n(\xi) d\xi + T_1 \int_{-1}^{+1} f_0(\xi) P_n(\xi) d\xi \right] \frac{2n+1}{2} \quad [3.19]$$

In the case of a surface temperature of the shape $T = T_0 + T_1 R \cos\theta$, we have:

$\alpha_0 = T_0, \quad \alpha_1 = T_1 R, \quad \alpha_n = 0$ for $n > 1$.

The temperature distribution inside the drop is then of the form $T = T_0 + T_1 r \cos\theta$ with:

$\int_{-1}^{+1} P_0(\xi) d\xi = 2$ and $\int_{-1}^{+1} P_n(\xi) d\xi = 0$ for n>1, and $\int_{-1}^{+1} \xi P_n(\xi) d\xi = \begin{cases} 2/3 & \text{for} \quad n = 1 \\ 0 & \text{for} \quad n > 1 \end{cases}$

We therefore solve the equation $\overline{\Delta}\overline{\Delta}\psi = 0$ with the interface boundary condition at $r = a$:

$$\tau_{r\theta} = \mu \left[r \frac{\partial}{\partial r}\left(\frac{v}{r}\right) + \frac{1}{r}\frac{\partial u}{\partial \theta} \right] = \frac{T_1}{a} \left| \frac{d\sigma}{dT} \right| f'_0(\xi) \sin\theta \quad [3.20]$$

and the flow cancellation conditions:

$$\int_0^{2\pi} \int_0^{a} v \sin\theta_0 \, r \, d\varphi \, dr = 0, \text{ and } \int_0^{2\pi} \int_0^{a} u_{r=r_0} r_0^2 \sin\theta_0 \, d\varphi \, d\theta = 0 \quad [3.21]$$

In the case of linear axial temperature: $T = T_0 + T_1 R \cos\theta$, we find the stream function:

$$\psi(r,\theta) = -\frac{T_1 R^3}{6\mu} \left| \frac{d\sigma}{dT} \right| \left[\left(\frac{r}{R}\right)^2 - \left(\frac{r}{R}\right)^4 \right] \sin^2\theta \quad [3.22]$$

which corresponds to the case of the Hill vortex of section 3.2.2.

Bauer also solves the problem in the case of an arbitrary stationary axial temperature field, or with a periodic time dependence.

3.3.2. *Thermo-capillary Hill vortex*

In the study by Bauer (1982), it is noted that the solution obtained in the case of a constant axial thermal gradient was a Hill vortex of the same axis.[3]

The result can be obtained directly as follows.

Consider a spherical drop of liquid whose surface is a phase separation of capillary tension σ. We assume that this capillary tension is a linear function of temperature:

$$\sigma = \sigma_0 + \sigma_T (T - T_0) \qquad [3.23]$$

The movement of the liquid is organized in a Hill vortex, but if we assume that the external fluid is perfect and incompressible, it is at rest.

We can ask ourselves the following question: *what temperature field is capable of generating a Hill vortex as a result of the movement of the surface of the drop by thermo-capillary effect?*

To answer this question, we establish the equilibrium conditions to be verified between the capillary forces and the viscous forces on the surface of the sphere.

At point *M*, the velocity vector is (Figures 3.2 and 3.3, with $U_\infty = V_0$ in the case of section 3.2):

$$\vec{\mathbf{v}}_i = \frac{3V_0}{2}\frac{y}{R^2}\left(y\vec{\mathbf{i}} - z\vec{\mathbf{k}}\right) = \frac{3V_0}{2}\left[\frac{r^2}{R^2}\vec{\mathbf{t}}\sin\theta + \left(\frac{r^2}{R^2} - 1\right)\vec{\mathbf{n}}\cos\theta\right] = \frac{3V_0}{2}\left(v_t\,\vec{\mathbf{t}} + v_n\,\vec{\mathbf{n}}\right) \qquad [3.24]$$

In *P*, i.e. for $r = R$, we have: $\vec{\mathbf{v}} = \vec{\mathbf{U}} = -2\alpha_i R^2 \sin\theta\,\vec{\mathbf{e}}_\theta = \frac{3V_0}{2}\vec{\mathbf{t}}\sin\theta$.

The balance of forces at a point on the capillary surface of the spherical drop involves the calculation of the tangential strain rate:

---

3 Unlike section 3.2, the spherical drop is not in the presence of a well-specified external flow. This is not of great importance if we neglect the interactions between the drop and the possible movements in its exterior.

$$\varepsilon_{tn} = \frac{1}{2}\left(r\frac{\partial (U_t/r)}{\partial r} + \frac{1}{r}\frac{\partial U_n}{\partial \theta} \right), \qquad [3.25]$$

which makes it possible[4] to express the tangential constraint $\tau_{tn} = 2\mu\varepsilon_{tn}$.

We have: $\frac{U_\theta}{r} = \frac{2\alpha_i}{r}\left(R^2 - 2r^2\right)\sin\theta,\ \ U_r = 2\alpha_i\left(r^2 - R^2\right)\cos\theta$, then:

$$\tau_{\theta r} = \tau_{r\theta} = 6\mu R\alpha_i \sin\theta. \qquad [3.26]$$

This tangential stress is equal to: $\frac{\partial\sigma}{\partial s} = \frac{1}{R}\frac{\partial\sigma}{\partial\theta} = \frac{\sigma_T}{R}\frac{\partial T}{\partial\theta}$ (Figure 3.6).

It follows that $\frac{\partial T}{\partial\theta} = \frac{4\mu R^2\alpha_i}{\sigma_T}\sin\theta$, either $T = T_0 - \frac{4\mu R^2\alpha_i}{\sigma_T}\cos\theta$ or:

$$T = T_0 - \frac{4\mu R\alpha_i}{\sigma_T} z \qquad [3.27]$$

We find a temperature field with a constant gradient.

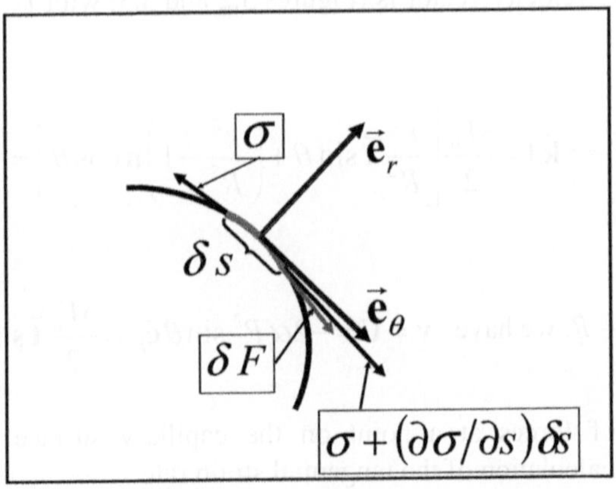

Figure 3.6. *Forces on the surface of the liquid. $\delta s = R\delta\theta$, F: viscous friction force. For a color version of this figure, see www.iste.co.uk/prudhomme/fluid3.zip*

4 Surface forces are calculated from the velocity field of the Hill vortex by introducing a viscosity, while this vortex is a perfect fluid flow. This is not contradictory if we assume that it is a local influence and that the body of the flow is very little modified.

We can therefore give the following result: *a spherical liquid drop of constant density, subjected to a uniform and constant temperature gradient* $\vec{\mathbf{G}} = \vec{\nabla}T$ *in an atmosphere at rest, is animated by the internal movement corresponding to the Hill vortex whose maximum intensity speed is oriented in the opposite direction to* $\vec{\mathbf{G}}$ *(Figure 3.7):* $U_{SG} = G\sigma_T R/2\mu$, *G being the thermal gradient* $G = dT/dz$.

$$U_{\max} = 3V_0/2 = -G\sigma_T R/2\mu \qquad [3.28]$$

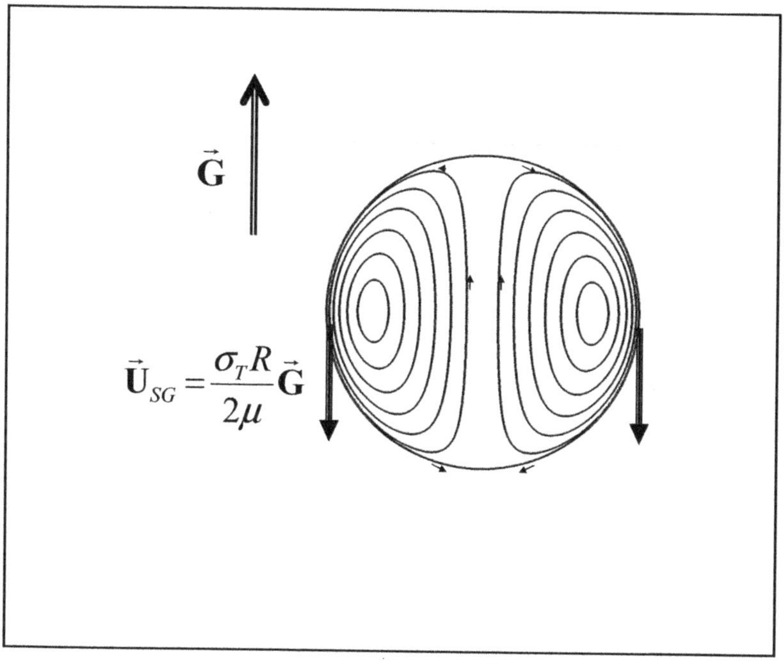

Figure 3.7. *Hill vortex generated by a thermal gradient*

3.3.3. *Other thermo-capillary flows*

As mentioned at the beginning of section 3.1, Bauer also deals with instabilities in cases other than the constant axial thermal gradient.

In his paper, Bauer (1985) studies the combined effects of Marangoni convection induced by a temperature gradient imposed on the free surface of a liquid sphere and natural convection originating from the residual microgravity field existing in an orbiting space laboratory (Lefebvre 2018). The case of a constant and linearly dependent axial residual gravity field has been considered, for which the Stokes equation in the Boussinesq approximation has been solved.

A dynamic Bond number derived from the ratio of the Grashof number, and the Reynolds number based on the Marangoni flow is introduced. It allows us to determine the predominance of the Marangoni effect if $Bo \rightarrow 0$ natural or convection if $Bo \rightarrow \infty$.

The combined effects of Marangoni and natural convection are then studied. Streamlines, radial and angular velocity distributions have been obtained analytically. On the contrary, isotherms are presented for different temperature distributions imposed on the free surface of the liquid sphere. The dynamic Bond number introduced is by definition:

$$\widetilde{B}o = \rho g \beta R^2 / |\sigma_T| \quad [3.29]$$

The main gravitational influence is due to the residual acceleration normal to the orbital path, in the plane of the orbit. It is created by the centrifugal acceleration of the space station in circular orbit and Newton's acceleration.

When g is constant, we write $g = g_0$, for variable g, we have $g = \Omega_0^2 R$ where the angular velocity of the center of mass around the center of the Earth is found.

Equations [3.12] to [3.14] still apply, but equation [3.15] is replaced by:

$$\overline{\Delta}\left[\overline{\Delta}\psi - \frac{1}{\nu}\frac{\partial \overline{\psi}}{\partial t}\right] = -\frac{3\Omega_0^2 \beta}{\nu}\left[r^2 \frac{\partial T}{\partial r}\sin\theta\cos\theta + r\frac{\partial T}{\partial \theta}\cos^2\theta\right]\sin\theta \quad [3.30]$$

where β is the coefficient of thermal expansion of the liquid.

We then use the recurrence formulas of the Legendre functions. The calculations lead to expressions for the stream function $\psi(r,\theta)$ and components $u(r,\theta)$, $v(r,\theta)$ of the velocity vector according to series developments of $P_n^0(\cos\theta)$ and $P_n^1(\cos\theta)$.

The constants involved in these expressions are to be determined according to the boundary conditions at the surface of the liquid sphere. In the absence of capillary effects, we obtain natural convection due to the residual gravity present at the location of the space station where the drop is located. We must then distinguish the case where the liquid is in a rigid container from the case of the free surface where $\widetilde{B}o \rightarrow \infty$ we can neglect thermo-capillary effects. We also resolve situations where the internal movements of the drop of thermo-capillary origin are significant, and we can thus predict the effects of residual gravity on these movements.

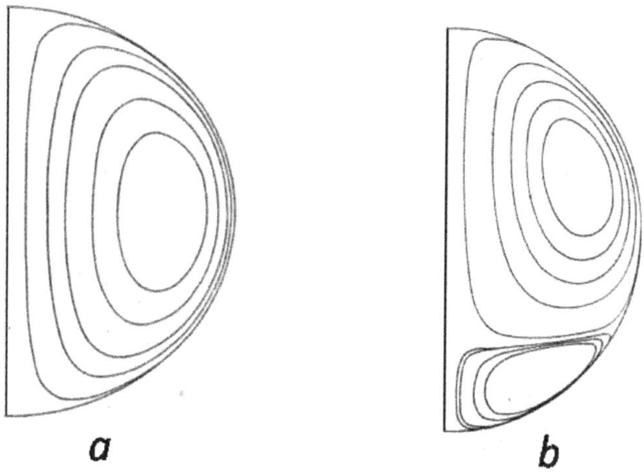

Figure 3.8. *Streamlines of vortex rings in the presence of a variable gravity field (Bauer 1985): a) for a linear and b) quadratic temperature field*

As for the gravity field, two situations are examined numerically: constant microgravity and linearly varying microgravity. As for the temperature field, the field is linear axisymmetric or mixed linear-quadratic $T = T_0 + T_1 \cos\theta + T_2 \cos^2\theta$.

Results of the calculations carried out show in Figure 3.8 the possibility of notable differences with Hill vortices. Two-ring configurations of current surfaces are observed (figures taken from the article of Bauer 1985) with a variable residual gravity field and quadratic temperature fields for $T_2/T_1 = 2$ with $\tilde{B}o = 0, 10, 100$.

Another figure from the cited paper also shows a case with two toric surfaces with a constant residual gravity field, the same quadratic temperature field and $\widetilde{B}o = 10$.

3.4. Conclusion

We presented the Hill vortex as a structure that can appear inside liquid drops in two circumstances: uniform external fluid flow at infinity and thermal gradient along an axial direction.

The physical hypotheses were:

– The sphericity of the drop of constant radius.

– The liquid: incompressible, dilatable or not, viscous but animated by a rotational movement of perfect fluid in a stationary regime.

– The external fluid: at rest or animated by a uniform movement at infinity, perfect irrotational or locally viscous.

– Concerning the external liquid–fluid interface:

- simple contact surface or sliding surface;

- identity of speeds: fluids-surface or only liquid-surface;

- with or without surface tension depending on the temperature.

We were concerned with situating the particular case of the Hill vortex as a particular case of analysis by Legendre function series.

Perspectives:

– On the Hill vortex, it will be necessary to conclude on the birth and dissipation of this vortex by providing characteristic times (Gharib et al. 1998; Chung 1982).

– On the Marangoni instability in a drop subjected to a spherically symmetric radial thermal field, we will have to refine our formulation of the problem in spherical coordinates using the articles of Hoefsloot et al. (1990, 1992)

– It would be interesting to study the effect of the thermo-capillary vortex on the evaporation of the drop (Shih and Megaridis 1995, 1996), especially in the presence of thermal radiation (Niazmand and Ambarsooz 2009) or acoustic excitation as we have done with other phenomena (Mauriot and Prud'homme 2014).

– Researchers are already interested in continuing investigations into the subject.[5]

3.5. Appendices

3.5.1. *Appendix 1: reminders on 3D irrotational flows of perfect incompressible fluids: source, sink, doublet, revolution flows*

3.5.1.1. *General (Germain 1962)*

Consider flows whose velocity vector derives from a potential ϕ. Let be $\vec{\mathbf{v}} = \vec{\nabla}\phi$ the velocity vector.

5 Notably at the University of Lomé (Togo) and at the University of Abomey-Calavi (Benin).

The continuity equation boils down to $\vec{\nabla}\cdot\vec{\mathbf{v}}=0$, from which we deduce: $\Delta\phi=0$. The function ϕ is therefore a harmonic function.

According to Bernoulli's second theorem, concerning irrotational flows (but which can be unsteady), we have, $\mathcal{V}$ designating the potential from which external forces at a distance are supposed to derive:

$$\mathcal{H}=\frac{\vec{\mathbf{v}}^2}{2}+\int\frac{dp}{\rho}+\mathcal{V}\text{, with: }\frac{\partial\varphi}{\partial t}+\mathcal{H}=C(t)$$

We give below three elementary examples of such flows: uniform flow, source and doublet. We can superimpose flows since the equations are linear.

3.5.1.1.1. Uniform flow

The uniform flow of speed U along the Ox axis has the speed potential:

$$\phi=U\,x_3 \tag{3.31}$$

3.5.1.1.2. Source and sink

A three-dimensional point source centered at O is defined by the potential:

$$\phi=-\frac{D}{4\pi r},\quad r^2=x_1^2+x_2^2+x_3^2 \tag{3.32}$$

The velocity vector is: $\vec{\mathbf{v}}=\vec{\nabla}\phi=\dfrac{D}{4\pi}\dfrac{\overrightarrow{\mathbf{OM}}}{r^3}$.

For a source, the flow rate D is positive: D > 0, whereas for a sink, it is negative: D < 0.

3.5.1.1.3. Doublet

Let a source and a sink be symmetrical with respect to the origin (Figure 3.9):

$$\phi=\frac{D}{4\pi}\left(\frac{1}{r_{A'}}-\frac{1}{r_A}\right),\quad \frac{1}{r_A}=\left[x_1^2+x_2^2+(x_3-a)^2\right]^{-\frac{1}{2}}\approx\frac{1}{r}\left[1+\frac{a\,x_3}{r^2}+\ldots\right]$$

We ask: $K = 2Da$, we obtain, when $a \to 0$ and $D \to \infty$:

$$\phi = -\frac{K}{4\pi}\frac{x_3}{r^3} = -\frac{K}{4\pi}\frac{\vec{\mathbf{e}}_3 \cdot \overrightarrow{\mathbf{OM}}}{r^3} \qquad [3.33]$$

This is an axis doublet $\vec{\mathbf{e}}_3$ and intensity *K*.

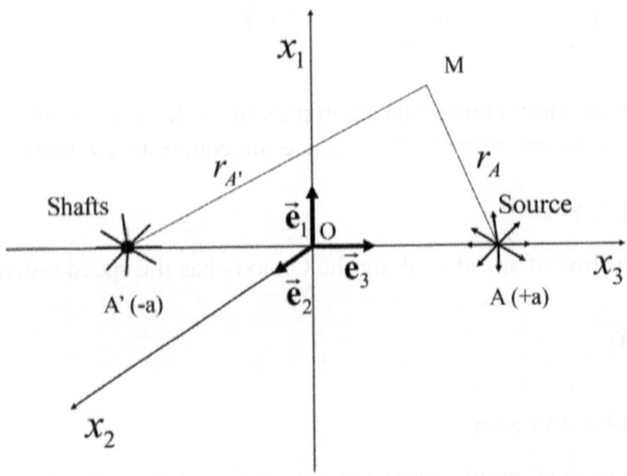

Figure 3.9. *Doublet in 3D space: we make the distance 2a between the well and the source tend towards zero to obtain a doublet located at the origin of the axes*

3.5.2. *Stationary irrotational revolution flows*

To study the flows of revolution, we choose a spherical coordinate system. r, θ, ϕ arranged as in Figure 3.10 relative to Cartesian coordinates $Ox_1\,x_2x_3$. By reason of symmetry, the flow is the same in any plane yOz passing through the axis $Oz = Ox_3$, and we call the unit vector $\vec{\mathbf{i}}$ normal to this plane in the direct direction.

There then exists $\psi(y,z)$ such as $\vec{\mathbf{v}} = -\frac{\vec{\mathbf{i}}}{y} \times \vec{\nabla}\psi$.

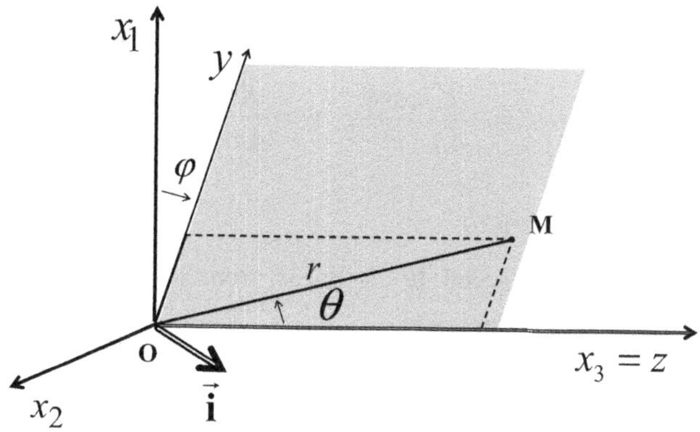

Figure 3.10. *The coordinate axis system in which the flow of revolution is studied. The spherical coordinates are r,θ,φ and we consider any plane yOz passing through the axis $Oz = Ox_3$, defined by an angle value φ*

In addition, since the flow studied is irrotational, we can write: $\vec{\mathbf{v}} = \vec{\nabla}\phi$. It follows that:

$$\frac{\partial\phi}{\partial z} = \frac{1}{y}\frac{\partial\psi}{\partial y}, \frac{\partial\phi}{\partial y} = -\frac{1}{y}\frac{\partial\psi}{\partial z} \qquad [3.34]$$

φ check: $\dfrac{\partial^2\phi}{\partial z^2} + \dfrac{1}{y}\dfrac{\partial}{\partial y}\left(y\dfrac{\partial\phi}{\partial y}\right) = 0$ and ψ check: $\dfrac{\partial^2\psi}{\partial z^2} + \dfrac{1}{y}\dfrac{\partial}{\partial y}\left(y\dfrac{\partial\psi}{\partial y}\right) = 0$.

3.5.3. *Examples*

3.5.3.1. *Stream functions of elementary examples*

The potentials and stream functions of the uniform flow of speed U_∞ verify:

$\phi = U_\infty z \Rightarrow \psi = \dfrac{U_\infty y^2}{2} + Const$, because $v_z = \psi_{,y}/y = U_\infty$ and $v_y = \psi_{,z}/y = 0$.

A *source flow* is defined as follows by the expression of its potential and its stream function:

$\phi = -\dfrac{D}{4\pi}\dfrac{1}{\left(y^2+z^2\right)^{1/2}} \Rightarrow \psi = -\dfrac{D}{4\pi}\dfrac{z}{\left(y^2+z^2\right)^{1/2}}$, or even: $\phi - \dfrac{D}{4\pi r}$ and $\dfrac{D}{4\pi}\cos\theta$

For an intensity *doublet K*, we have:

$$\phi = -\frac{K}{4\pi}\frac{z}{\left(y^2+z^2\right)^{3/2}} = -\frac{K}{4\pi}\frac{\cos\theta}{r^2}, \quad \psi = \frac{K}{4\pi}\frac{y^2}{\left(y^2+z^2\right)^{3/2}} = \frac{K}{4\pi}\frac{\sin^2\theta}{r}$$

3.5.3.2. *Flow around a sphere*

The system verified by ϕ and ψ is linear, from which it follows that linear combinations of solutions are possible.

For example, the superposition of a uniform flow and a doublet gives the flow around a sphere:

$$\phi = U_\infty z - \frac{K}{4\pi}\frac{z}{r^3}, \quad \psi = \frac{U_\infty}{2}y^2 + \frac{K}{4\pi}\frac{y^2}{r^3}, \text{ either, with } K = -2\pi U_\infty R^3:$$

$$\phi = U_\infty z\left(1+\frac{R^3}{2r^3}\right), \quad \psi = \frac{U_\infty}{2}y^2\left(1-\frac{R^3}{r^3}\right) \quad [3.35]$$

The resulting velocity fields are:

$$\begin{aligned} v_z &= \phi_{,z} = \psi_{,y}/y = U_\infty\left[1+\frac{R^3}{2r^5}\left(y^2-2z^2\right)\right] \\ v_y &= \phi_{,y} = -\psi_{,z}/y = -\frac{3}{2}U_\infty R^3\frac{yz}{r^5} \end{aligned} \quad [3.36]$$

On the surface of the sphere at $r = R$, we can write:

$$U_x = \frac{3}{2}U_\infty y^2/R^2, \; U_y = -\frac{3}{2}U_\infty xy/R^2 \Rightarrow \frac{U_y}{U_x} = -\frac{x}{y}.$$

The velocity vector $\vec{\mathbf{v}}$ is indeed tangent to the sphere. On the contrary, $v = |\vec{\mathbf{v}}| = \frac{3}{2}U_\infty|\sin\theta|$.

Bernoulli's theorem gives:

$$\frac{p}{\rho} + \frac{\mathbf{v}^2}{2} = \frac{p_\infty}{\rho} + \frac{U_\infty{}^2}{2}, \qquad p = p_\infty + \rho\frac{U_\infty{}^2}{2}\left(1-\frac{9}{4}\sin^2\theta\right)$$

The resultant of the forces exerted on the sphere is obtained by integration of $-p\vec{\mathbf{n}}$, where $\vec{\mathbf{n}}$ is the external normal to the surface of the sphere. For reasons of symmetry, it is therefore zero. The resultant of the forces exerted by a fluid flow on a stationary sphere would therefore be zero.

This is *d'Alembert's paradox*. In reality, the resultant is not zero, because friction forces are exerted on the surface of the sphere due to viscosity, which is not considered in the previous calculation.

3.5.4. *Appendix 2: rotational flows of perfect incompressible fluids*

Here, we no longer assume the irrationality of the flow.

3.5.4.1. *Equation of vortex transport*

If the mass forces derive from a potential, then the rotation rate vector (or "vortex") $\vec{\boldsymbol{\omega}} = \frac{1}{2}\overrightarrow{\mathbf{rot}}\,\vec{\mathbf{v}}$, nonzero, is transported by the movement. We have in fact:

$$\frac{d\vec{\boldsymbol{\omega}}}{dt} = \vec{\nabla} \otimes \vec{\mathbf{v}} \cdot \vec{\boldsymbol{\omega}} \qquad [3.37]$$

We thus find a property of the theory of deformations, according to which, for any infinitesimal material vector: $\frac{d\,\delta\vec{\mathbf{M}}}{dt} = \vec{\nabla} \otimes \vec{\mathbf{v}} \cdot \delta\vec{\mathbf{M}}$. This property applies here to the vortex.

This is demonstrated as follows: the equation of the quantity of movement of a perfect fluid is written as:

$\frac{\partial \vec{\mathbf{v}}}{\partial t} + \overrightarrow{\mathbf{rot}}\,\vec{\mathbf{v}} \times \vec{\mathbf{v}} + \vec{\nabla}\mathcal{H} = \vec{\mathbf{0}}$, with: $\mathcal{H} = \frac{\vec{\mathbf{v}}^2}{2} + \frac{p}{\rho} + V$ incompressible. Taking the rotational, we obtain: $\frac{\partial \vec{\boldsymbol{\omega}}}{\partial t} + \overrightarrow{\mathbf{rot}}(\vec{\boldsymbol{\omega}} \times \vec{\mathbf{v}}) = \vec{\mathbf{0}}$, which is equivalent to equation [3.34].

If at a given t we know $\vec{\omega}$, we deduce from this $\vec{v}$ and the equation governing the evolution of $\vec{\omega}$ will allow, if we know how to solve it, to determine the speed field.

Equations for finding the velocity vector $\vec{v}$:

$\vec{\nabla} \cdot \vec{v} = 0$, $\overrightarrow{\mathbf{rot}}\, \vec{v} = 2\vec{\omega}$ (assumed known). If $\vec{\omega}$ is nonzero in a bounded domain (and zero elsewhere), we can form a particular solution (canonical flow) generated by $\vec{\omega}$:

$$\vec{v}^{(2)}(M) = -\frac{1}{2\pi}\int \frac{\overrightarrow{MM'} \times \vec{\omega}(M')}{\left|\overrightarrow{MM'}\right|^3} d\mathcal{V}(M')$$

An irrotational flow can be added to this solution $\vec{v}^{(e)}$ (irrotational entrainment flow). This is possible in particular if $\vec{\omega} \cdot \vec{n} = 0$ at the frontiers.

3.5.4.2. *Plane rotational movements*

Let $\vec{\omega} = \omega \vec{\mathbf{k}}$, $\dfrac{d\omega}{dt} = 0$ (indeed: $\vec{\mathbf{k}}\, d\omega/dt = \omega \vec{\nabla} \otimes \vec{v} \cdot \vec{\mathbf{k}}$ but $\vec{\nabla} \otimes \vec{v}$ is in the (x, y) plane. $\rightarrow \dfrac{d\omega}{dt} = 0$) (Figure 3.10).

Stream function Ψ: we have $\vec{v} = -\vec{\mathbf{k}} \times \vec{\nabla}\psi = \overrightarrow{\mathbf{rot}}\left(\psi \vec{\mathbf{k}}\right)$.

NOTE.– *In the general case,* $\vec{\nabla} \cdot \vec{v} = 0 \rightarrow \vec{v} = \overrightarrow{\mathbf{rot}}\left(\vec{\mathbf{W}}\right)$ *and we can choose* $\vec{\mathbf{W}}$ *such as* $\vec{\nabla} \cdot \vec{\mathbf{W}} = 0$*, then:* $\vec{\nabla} \times \left(\vec{\nabla} \times \vec{\mathbf{W}}\right) = \vec{\nabla}\left(\vec{\nabla} \cdot \vec{\mathbf{W}}\right) - \Delta \vec{\mathbf{W}} \rightarrow \Delta \vec{\mathbf{W}} = -2\vec{\omega}$. $\psi \vec{\mathbf{k}}$ *is an individual* $\vec{\mathbf{W}}$*; therefore,* $2\vec{\omega} = -\vec{\nabla}\psi$.

EXAMPLE.– *The Rankine vortex. This vortex results from the superposition of the movement of a solid, non-deformable disk of radius a with the angular speed of rotation Ω and the movement of an irrotational vortex of intensity Γ outside this disk. With a normal third direction to the plane, this example is representative of the drain vortex.*

The connection condition at $r = a$ is then $\Omega = \Gamma/2\pi$. The field of the current function is as follows:

$$\begin{cases} \psi = -(\Gamma/4\pi)\left(r^2/a^2 - 1\right), \; r \le a \\ \psi = -(\Gamma/2\pi)\ln r/a, \; r \ge a \end{cases}$$

3.5.5. *Appendix 3: details of calculations valid inside the sphere*

This appendix relates to section 3.2.2. The solution in the rotational zone ($r \le R$) is written as: $\psi = \alpha r^2\left(R^2 - y^2 - z^2\right)$ and $\omega = 5\alpha r$, with α, R constants, $\alpha = -3U_\infty/4R^2$. We have $\psi/\alpha \ge 0$. By posing $r^2 = y^2 + z^2$ and $z/y = \text{tg}\,\varphi$, it becomes: $\cos^2\varphi = \dfrac{\psi}{\alpha}\dfrac{1}{r^2\left(R^2 - r^2\right)}$. The streamlines are shown in Figure 3.4 in reduced coordinates. For $\varphi = 0, \cos^2\varphi = 1, \alpha r^2\left(R^2 - r^2\right) = \psi$, which gives two values of r^2, (therefore of r), provided that $\dfrac{r^4}{4} - \dfrac{\psi}{\alpha} \ge 0 : r^2 = \dfrac{R^2}{2} \pm \sqrt{\dfrac{R^4}{4} - \dfrac{\psi}{\alpha}} \ge 0$. Let $r_m^2 = \dfrac{R^2}{2} - \sqrt{\dfrac{R^4}{4} - \dfrac{\psi}{\alpha}} \ge 0$ and $r_M^2 = \dfrac{2R^2}{2} + \sqrt{\dfrac{R^4}{4} - \dfrac{\psi}{\alpha}} \ge 0$.

– If $\psi = 0$, we find $r_m = R$, $r_M = 0$, which corresponds to the circle with center O, therefore to the sphere, which is a current surface, and to its center. In other cases, we have the condition: $0 \le \psi/\alpha \le R^4/4$.

– If $\psi = \alpha R^4/4$, the streamline narrows to the point $\varphi = 0, r = R/\sqrt{2}$.

– Maximum of $\cos^2\varphi = \dfrac{\psi}{\alpha}\dfrac{1}{r^2\left(R^2 - r^2\right)}$:

$\cos^2\phi$ is maximum, with data value for ψ, for $r = R\sqrt{2}/2$, which gives the quarter circle $\overline{X}_M$ of the Figure 3.1(b). For $\psi = 0$, $\phi_M = \pi/2$. For $\psi/\alpha = R^2/4$, $\phi_M = 0$.

– Minimum speed: the modulus of the speed vector is equal to $|U| = 2\alpha\left[y^2 z^2 + \left(R^2 - 2y^2 - z^2\right)^2\right]^{1/2}$. Consider the quantity $\overline{K} = U^2/4\alpha^2 R^4$, square of the reduced velocity, which is also a reduced kinetic energy. Its expression as a function of $\bar{y} = y/R$, $\bar{z} = z/R$ is: $\bar{K} = \bar{y}^2 - \frac{3\bar{\psi}}{4} + \frac{\bar{\psi}^2}{16}\frac{1}{\bar{y}^4}$ with $\bar{\psi} = 4\psi/\alpha R^4$, reduced current function. The minimum of $\bar{K}$ (i.e. of U) is obtained for $\bar{y}^2 = \bar{\psi}^{2/3}/2$, which gives the geometric locus of equation $\bar{z}_U = 1 - \bar{y}^2 - \bar{y}/\sqrt{2}$ shown in Figure 3.1(b).

$\bar{K}_m = 0.75\left(\bar{\psi}^{2/3} - \bar{\psi}\right)$ is shown in Figure 3.6.

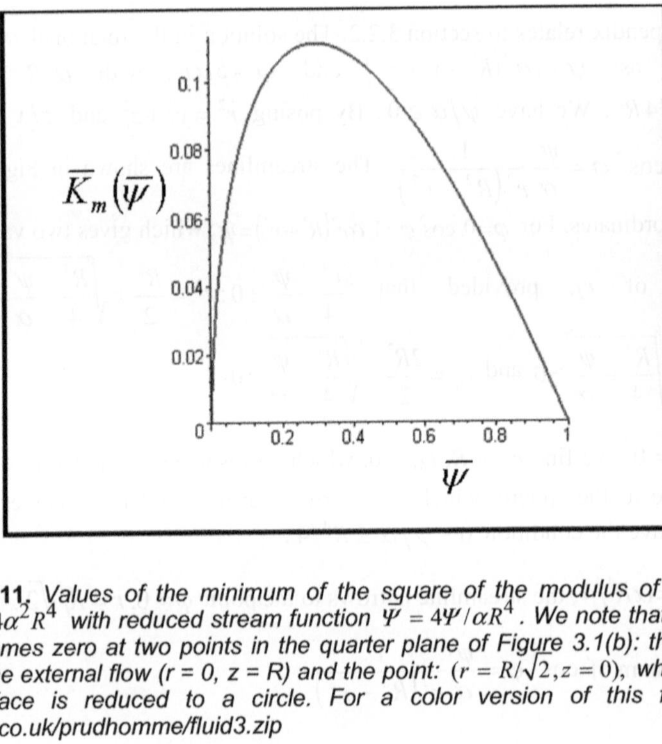

Figure 3.11. *Values of the minimum of the square of the modulus of the speed $\bar{K} = U_m^2/4\alpha^2R^4$ with reduced stream function $\bar{\Psi} = 4\Psi/\alpha R^4$. We note that the speed only becomes zero at two points in the quarter plane of Figure 3.1(b): the stopping point of the external flow (r = 0, z = R) and the point: $(r = R/\sqrt{2}, z = 0)$, where the Hill toric surface is reduced to a circle. For a color version of this figure, see www.iste.co.uk/prudhomme/fluid3.zip*

We note that the speed only becomes zero at two points in the quarter plane of Figure 3.1(b): the stopping point of the external flow and the point, where the Hill toric surface is reduced to a circle.

Circulation of the velocity vector

We know that if (C) is a closed contour and (S) a piecewise regular surface, open and resting on the contour (C), if $\vec{u}$ is a regular function of the coordinates, then:

$$\oint_C \vec{u} \cdot \vec{t}\, dl = \iint_S \vec{\nabla} \times \vec{u} \cdot \vec{n}\, dS = 2 \iint_S \vec{\omega} \cdot \vec{n}\, dS$$

This circulation can be easily calculated on the streamline consisting of the semicircle of radius $R, y \geq 0$ and diameter $y = 0, -R \leq z \leq +R$. By separating the variables, we obtain:

$$\Gamma = 20\alpha \int_0^R r^2 dr \int_{\pi/2}^0 \cos\varphi\, d\varphi = -\frac{20}{3}\alpha R^3.$$

3.5.6. *Appendix 4: Legendre polynomials and spherical harmonics*

3.5.6.1. *Reminder on Legendre polynomials*

The *Legendre polynomials*[6] are the simplest example of a sequence of orthogonal polynomials[7]. They are polynomial solutions $P_n(x)$ of the Legendre differential equation:

$$\frac{d}{dx}\left[\left(1-x^2\right)\frac{d}{dx}P_n n(x)\right] + n(n+1)P_n(x) = 0$$

in the special case where the parameter n is an integer.

The Legendre polynomials are defined only for x ∈ [−1; 1] since the points x = ±1 are regular singular points of this differential equation[1].

Representative curves of the first six Legendre polynomials: $P_0(x) = 1$, $P_1(x) = x$, $P_2(x) = (3x^2-1)/2$, $P_3(x) = (5x^3-3x)/2$, $P_4(x) = (35x^4-30x^2+3)/8$, $P_5(x) = (63x^5-70x^3+15x)/8$ (seen in Figure 3.1).

6 See J. Bass' Mathematics Course (1968).

7 In mathematics, a *sequence of orthogonal polynomials* is an infinite sequence of polynomials $p_0(x)$, $p_1(x)$, $p_2(x)$ ... with real coefficients, in which each $p_n(x)$ is of degree n, and such that the polynomials of the sequence are orthogonal two by two for a given definition of the scalar product of functions. If $\langle p_n, p_m \rangle = 0$ for n different from m, the polynomials are therefore orthogonal. The simplest scalar product of functions is the integral of the product of these functions, on a bounded interval: $\langle f, g \rangle = \int_a^b W(x) f(x) g(x) dx$, where $W(x)$ is a weight function.

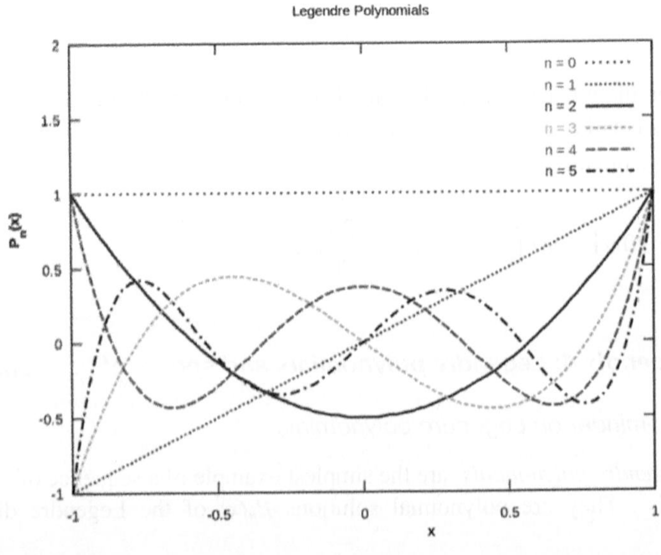

Figure 3.12. *The simplest orthogonal polynomials are the Legendre polynomials for which the orthogonality interval is]-1, 1 (and the weight function W(x) is the constant function of value 1). Representative curves of the first six Legendre polynomials: $P_0(x)$ = 1, $P_1(x)$ = x, $P_2(x)$ = (3x2-1)/2, $P_3(x)$ = (5x3-3x)/2, $P_4(x)$ = (35x4-30x2+3)/8, $P_5(x)$ = (63x5-70x3+15x)/8. For a color version of this figure, see www.iste.co.uk/prudhomme/fluid3.zip*

These orthogonal polynomials have many applications both in mathematics, for example for the decomposition of a function into a series of Legendre polynomials, and in physics, where the Legendre equation appears naturally when solving the Laplace or Helmholtz equations in spherical coordinates.

3.5.6.2. *The spherical harmonics*[8]

Let the operator be $\mathcal{L}^2 = -\frac{1}{\sin\theta}\frac{\partial}{\partial\theta}\left(\sin\theta\frac{\partial}{\partial\theta}\right) - \frac{1}{\sin^2\theta}\frac{\partial^2}{\partial\varphi^2}$. The spherical harmonics are such that $\mathcal{L}^2 Y_l^m(\theta,\varphi) = -l(l+1)Y_l^m(\theta,\varphi)$.

We then show that: $Y_l^m(\theta,\varphi) = P_{l,m}(\cos\theta)e^{im\varphi}$, where $P_{l,m}(\cos\theta)$ is a Legendre polynomial. The eigenvalue equation becomes a second-order

8 See, for example, the Wikipedia site on this question.

differential equation for the function $P_l^m(\cos\theta)$:

$$-\frac{1}{\sin\theta}\frac{d}{d\theta}\left(\sin\theta\frac{dP_{l,m}(\cos\theta)}{d\theta}\right)+\frac{m^2}{\sin^2\theta}P_{l,m}(\cos\theta)=E_{l,m}P_{l,m}(\cos\theta)\cdot$$

The change of variable $x=\cos\theta$ leads to Legendre's generalized differential equation:

$$-\frac{d}{dx}\left(\left(1-x^2\right)\frac{dP_{l,m}(x)}{d\theta}\right)+\frac{m^2}{1-x^2}P_{l,m}(x)=E_{l,m}P_{l,m}(x)$$

The eigenvalues of this equation are independent of m:

$$E_{l,m}=l(l+1)$$

The eigenfunctions $P_{l,m}(x)$ are constructed from the Legendre polynomials which are eigenfunctions of the Legendre ordinary differential equation, corresponding to the case $m = 0$:

$$-\frac{d}{dx}\left[\left(1-x^2\right)\frac{d}{dx}P_l(x)\right]=l(l+1)P_l(x)=0$$

We have the generating formula of Olinde Rodrigues:

$$P_l(x)=\frac{1}{2^l l!}\frac{d^l}{dx^l}\left(x^2-1\right)^l.$$

– For non-integer l, we find solutions tending to infinity when x tends to (-1).

– The only admissible solutions are therefore those defined for integer l.

– We find: $P_l(x)=\frac{(-1)^m}{2^l l!}\left(1-x^2\right)^{m/2}\frac{d^{l+m}}{dx^{l+m}}\left(x^2-1\right)^l$.

– P_l^m and P_l^{-m} are two eigenfunctions corresponding to the same eigenvalue $l(l+1)$, but in reality, they are proportional: $P_{l,(-m)}(x)=\frac{(n-m)!}{(n+m)!}P_{l,m}(x)$.

– Since spherical harmonics form an orthogonal basis on the unit sphere, any continuous function $f(\theta,\varphi)$ decomposes into a series of spherical harmonics:

$f(\theta,\varphi)=\sum_{l=0}^{l=+\infty}\sum_{m=-l}^{m=+l} C_l^m Y_l^m(\theta,\varphi)$, where l and m are integer indices, C_l^m is a coefficient constant which in mathematics often takes the name of generalized Fourier coefficient relative to this base (the development in spherical harmonics is the equivalent, applied to angular functions, of the development in Fourier series for periodic functions).

– Y_l^m is the real part of a complex function $\underline{Y_l^m}$: $Y_l^m(\theta,\varphi)=\Re\left(\underline{Y_l^m(\theta,\varphi)}\right)$. $\underline{Y_l^m}$ is called the "Legendre associated function" and is defined by $\underline{Y_l^m}(\theta,\varphi)=\sqrt{\frac{2(l-m)!}{(l+m)!}}P_{l,m}(\cos\theta)e^{im\varphi}$, where $P_{l,m}(\cos\theta)$ is the Legendre polynomial $P_{l,m}(X)=\frac{(-1)^m}{2^l l!}\left(1-X^2\right)^{m/2}\frac{\partial^{m+l}}{\partial X^{m+l}}\left[\left(X^2-1\right)^l\right]$. For example, we have the functions in Figure 3.1. The functions $Y_l^m(\theta,\varphi)$ exhibit more and more symmetries as l increases (except when $l=0$, since Y_0^0 is a constant function and therefore describes a sphere).

– Generalized spherical harmonics, defined on the sphere S_3.

Normalization of spherical harmonics leads to the final expression:

$$Y_l^m(\theta,\varphi)=\sqrt{\frac{(2l+1)(l-m)!}{4\pi(l+m)!}}P_{l,|m|}\cos\theta\, e^{im\varphi}$$

3.6. References

Abramzon, B. and Sirignano, W.A. (1989). Droplet vaporization model for spray combustion calculations. *Int. J. Heat Mass Transfer*, 32(9), 1605–1618. doi: 10.1016/0017-9310(89)90043-4.

Bauer, H.F. (1982). Marangoni convection in a freely floating liquid sphere due to axial temperature field. *Ing. Arch.*, 52, 263–273.

Bauer, H.F. (1985). Combined residual natural and Marangoni convection in a liquid sphere subjected to a constant and variable micro-gravity field. *Zamm Z. angew. Math. Mech.*, 65(10), 461–470.

Bauer, H.F. and Eidel, W. (1987). Marangoni convection in a spherical liquid system. *Acta Astronautica*, 15(5), 275–290.

Chung, J.N. (1982). The motion of particles inside a droplet. *Trans. ASME*, 104(3), 438–445.

Germain, P. (1962). *Mécanique des milieux continus*. Masson, Paris.

Germain, P. (1986). *Mécanique*. Ellipses, Palaiseau.

Gharib, M., Rambod, E., Shariff, K. (1998). A universal time scale for vortex ring formation. *J. Fluid Mech.*, 360, 121–140.

Hoefsloot, H.C.J., Hoostraten, H.W., Janssen, L.P.B.M. (1990). Marangoni instability in a confined between two concentric spherical surfaces liquid layer under zero-gravity conditions. *Appl. Sci. Res.*, 47, 357–377.

Hoefsloot, H.C.J., Hoostraten, H.W., Janssen, L.P.B.M., Knobbe, J.W. (1992). Grows factors for Marangoni instability in a spherical liquid layer under zero-gravity conditions. *App. Sci. Res.*, 49(2), 161–173.

Lamb, H. (1945). *Hydrodynamics*. Cambridge University Press, Cambridge.

Lefebvre, D. (2018). Microgravité dans l'ISS [Online]. Available at: http://www.tangentex.com/Microgravite.htm.

Mauriot, Y. and Prud'homme, R. (2014). Assessment of evaporation equilibrium and stability concerning an acoustically excited drop in combustion products. *Comptes rendus mécanique*, 342(4), 240–253. doi: 10.1016/j.crme.2014.02.004.

Niazmand, H. and Ambarsooz, M. (2009). Effect of slip and Marangoni convection on single fuel droplet heat-up in the presence of thermal radiation. In *6th International Chemical Engineering Congress & Exhibition (ICHEC)*, Kish Island.

Prud'homme, R. (2012). *Flows and Chemical Reactions*. ISTE Ltd, London and John Wiley & Sons, New York.

Shih, A.T. and Megaridis, C.M. (1995). Suspended droplet evaporation modeling in a laminar convective environment. *Combust. Flame*, 102(3), 256–270.

Shih, A.T. and Megaridis, C.M. (1996). Thermocapillary flow effects on convective droplet evaporation. *Int. J. Heat Mass Transfer*, 39(2), 247–257.

4

The Fire of Weapons

This chapter gives a brief overview of combustion-based weapons and their effects. Flamethrowers and thermobaric weapons are discussed, as well as bombs such as napalm, high explosives and phosphorus bombs. Missiles, space armies and weapons pollution are also discussed.

4.1. Introduction

A draft chapter entitled "What is combustion", written as a proposal for a series, elicited a reaction from the editor overseeing its progress. The editor felt that the chapter's summary was too neutral and not engaging enough for readers, and therefore sought to enhance it.

To highlight the importance and interest of the topic, the following text was added to the OpenScience journal[1]:

> The discovery of fire marked the dawn of humanity. It was the scientific mastery of combustion that enabled humans to reach the moon, or more modestly, to make your car engine run efficiently. Who hasn't been captivated by fireworks, a simple candle, or a wood fire? Numerous physical and chemical phenomena are at play, which this article will help you uncover.

Chapter written by Roger PRUD'HOMME.

1 Prud'homme, R. (2019). Notions de base sur la combustion. *Thermodynamique des interfaces et mécanique des fluides*, 1, 3. doi: 10.21494/ISTE.OP.2019.0376.

The quoted chapter was translated into English for a book[2].

However, this portrayal of the discovery of fire[3] and its consequences may seem overly positive when considering the destructive use of *fire of weapons*[4], warfare and fires intentionally set for destruction. Sophisticated and inhumane weapons are often used against civilians, causing death and immense suffering. While some argue that this is part of deterrence and self-defense, there are also crimes against humanity – a concept that emerged after World War II – that continue to shape our reality. The struggles for peace, like those for women's rights to freedom and life, are long and arduous battles.

Moreover, many discoveries beyond fire have led to research for various purposes, not all of which have contributed to human progress. It is the responsibility of societies to ensure that discoveries genuinely benefit people.

It is beneficial to present combustion interfaces from another perspective of human activity, similar to how we address the issue of pollution. It is important to remember that significant discoveries have not always been used for peaceful purposes. This was true for the discovery of fire, its mastery and its use by humans. Human lives have been (and continue to be) lost as a result[5], with the added issue of pollution often overlooked.

The intention here is not to question the necessity of national defense, which remains crucial in the absence of global peace but to highlight the dangerous consequences of aggressive warfare. This indirectly encourages the reader to strive for a peaceful world.

This chapter primarily presents the main "flame weapons", which use flames to directly destroy installations or people. It then discusses missiles, the Space Army and pollution. The conclusion offers pacifist suggestions, followed by two appendices. Appendix 1 (section 4.7.1) provides reflections on the chapter's meaning. Appendix 2 (section 4.7.2) covers conventional firearms and includes additional information on napalm, white phosphorus and the Geneva Conventions.

2 Prud'homme, R. (2021). Basics on interfaces in combustion. In *Fluid Mechanics at Interfaces 1: Methods and Diversity*, Prud'homme, R. and Vincent, S. (eds). ISTE Ltd, London and John Wiley & Sons, New York.

3 In fact, it is the mastery of fire that dates back some 400,000 years.

4 Jas Van Driel, FESAC advisor, is the author of a *Brief History of Firearms*, published on the website: https://www.armes-ufa.com/spip.php?article39.

5 There are the consequences of armed conflicts, civil wars and the use of weapons in peacetime (in the United States alone, according to the Gun Violence Archive website, firearms will cause more than 43,000 deaths in 2020, including suicides).

4.2. The "flame" weapon

4.2.1. *The flamethrower*

The *flamethrower* is a land-based weapon designed to project flaming liquid. It can be either carried by an infantryman or integrated into a tank's armament.

Its lethal effect is achieved by either directly spraying the victim with the flaming liquid, exposing the victim to intense heat or consuming oxygen in an enclosed space.

The weapon is also intended to act as a deterrent, instilling fear through the prospect of being burned alive.

In addition, it can be used to set fire to buildings or vehicles.

Figure 4.1. *A 12th-century image showing the use of wildfire by the Byzantine navy against the rebel fleet of Thomas the Slav between 820 and 823[6]. For a color version of this figure, see www.iste.co.uk/prudhomme/fluid3.zip*

This weapon has ancient predecessors, one of the oldest being Greek fire (Figure 4.1). *Greek fire* (from the Latin *græcus*) is a flammable mixture that continues to burn even when in contact with water. It was used in ancient and medieval times to create incendiary devices for sieges and naval battles. Its invention dates back to the late 7th century and is attributed to Callinicus of Heliopolis, an architect who sought refuge in Constantinople.

6 See: https://en.wikipedia.org/wiki/Thomas_the_Slav.

An example of a modern flamethrower is shown in Figure 4.2.

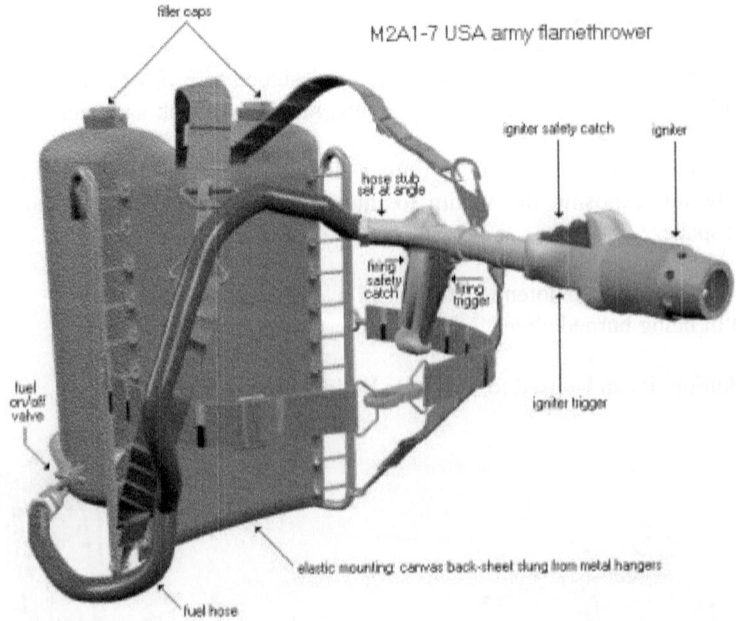

Figure 4.2. *The US Army's M2A1-7 flamethrower. For a color version of this figure, see www.iste.co.uk/prudhomme/fluid3.zip*

The flamethrower was used in several conflicts during the Cold War, including Korea and Vietnam, where American forces deployed the latest version of their portable flamethrower, the M9A1, derived from the M2A1.

The last instance of the French Army using the weapon was in 1988, during the Ouvéa hostage crisis in New Caledonia[7], to neutralize a machine gun.

4.2.2. *The thermobaric weapon*

Discontinued since the 1970s/1980s, flamethrowers have been replaced by thermobaric weapons in most armies (Figure 4.3).

7 The Ouvéa hostage taking, which lasted from April 22 to May 5, 1988, was the culmination of a series of violent events on the island of Ouvéa (New Caledonia) that left a lasting mark on the territory. See: https://en.wikipedia.org/wiki/Ouv%C3%A9a_cave_hostage_taking.

Figure 4.3. *Firing of a US Navy fuel–air explosive against a target ship, the USS McNulty, 1972*[8]*. For a color version of this figure, see www.iste.co.uk/prudhomme/fluid3.zip*

A *thermobaric weapon* is a conventional explosive weapon that combines thermal, shockwave and vacuum effects.

4.3. The bombs

4.3.1. *Napalm*

Napalm[9], a gelled form of petrol invented in 1942, is typically used in incendiary bombs. This substance is engineered to burn at a specific temperature and adhere to objects and people (Figures 4.4 and 4.5). In 1980, its use against civilians was prohibited by a United Nations convention and international humanitarian law, based on the Geneva Conventions, which banned weapons that cannot differentiate between civilians and combatants.

8 A detailed description can be found at https://en.wikipedia.org/wiki/Thermobaric_weapon where several thermobaric explosive weapons are listed.

9 See: https://en.wikipedia.org/wiki/Napalm.

Figure 4.4. *Depicts the burns suffered by the "little napalm girl" and treated 40 years later. This photograph of the granddaughter, who was burned during a pro-American napalm attack in the Vietnam War on June 8, 1972, gained international attention. More than four decades later, Kim Phuc is finally undergoing laser treatment to heal her scars. Artist: Nick Ut (Vietnamese, born 1951), who captured the Pulitzer Prize-winning photograph of children fleeing a napalm attack in South Vietnam, 1973 (source: Google[10])*

Figure 4.5. *French napalm bombing of Phu Nho Quan during the Indochina War, 1954*

10 See: https://en.wikipedia.org/wiki/Nick_Ut. In the background, dense black smoke rises from the village of Trang Bang near Saigon, as South Vietnamese planes drop napalm on a Buddhist pagoda, wrongly assuming it has been taken over by Vietcong fighters.

4.3.2. *Explosive devices*

The overwhelming power of explosive devices includes the thunder of the shells, the noise and the fire. During World War I, over 856 million shells were fired, resulting in devastated landscapes and traumatized individuals (source: Mission centenaire 14–18[11]).

World War II witnessed vast destruction from shells and bombs, leading to both chemical and nuclear explosions.

A *bomb* is an explosive device (Figure 4.6), usually consisting of a container filled with explosive material designed to cause destruction when detonated. Since the 20th century, the term has primarily been used to describe explosive devices launched from aircraft.

When a bomb detonates, it creates a pressure wavefront. The speed of this wavefront determines the classification of the explosive.

There are two main types of explosives:

– *Powder or blasting explosives (deflagration regime)*: these powders or solid propellants can also be used to propel projectiles or rockets.

– *High explosives (detonation mode)*: the most powerful known explosive (octanitrocubane) reaches a detonation velocity of 10,100 m/s. Other high explosives include nitro and nitrate groups, organic peroxides, chlorates and perchlorates, nitrogen halides, azides and fulminates.

The difference between deflagration and detonation is not always straightforward. Under certain conditions, an explosive typically meant to deflagrate can detonate and vice versa. Powders are designed for deflagration, which involves subsonic self-combustion (the chemical reaction propagates at a speed of 10 to 400 m/s within the explosive itself).

Progressive explosives fall between powders and high explosives, following the supersonic detonation regime (2,000–3,500 m/s). Brittle explosives detonate at speeds ranging from 4,000 to 9,000 m/s (source: Explosive[12]).

11 See: http://centenaire.org/fr/autour-de-la-grande-guerre/peinture/objets/explosions.

12 See: https://en.wikipedia.org/wiki/Explosive.

Figure 4.6. *Use of explosives by engineers during World War I (source: original uploaded by Boldair on French Wikipedia)*

4.3.3. *Phosphorus bombs*

The impact of using this weapon is devastating. The glowing particles of white phosphorus – nicknamed Willy Pete[13] by British soldiers during World War II and US soldiers in Vietnam – penetrate deeply into the skin, dissolving epidermis, flesh and bone. White phosphorus inflicts multiple chemical burns, which can continue to burn within the body, even without external oxygen:

> Usually, when a patient has a burn, we know how to treat it and, most importantly, there is no deterioration, explains Dr. Nafez Abou Shaaban, head of the burns department at Shifa Hospital. In this case, not only was treatment impossible, but the wound kept expanding, and after a few hours, white smoke started to emerge from it. The only solution was to get the patient to the operating room as soon as possible[14].

These burns are often second and third degree.

In its assault on Gaza, the Israeli military deployed white phosphorus munitions. During the UNWRA school bombing in Beit Lahia, Gaza, as part of the 2008–2009

13 White Phosphorus: WP.

14 *White Phosphorus Still Burning in Gaza Hospital* (article by Pierre Barbancey in L'Humanité, January 22, 2009).

"Operation Cast Lead", the use of white phosphorus by the Israeli forces resulted in the deaths of two children, the maiming of their mother and injuries to dozens of individuals, including children.

Those who inhaled the white smoke laced with toxic particles are likely to suffer long-term effects. This weapon had previously been used by the United States in Vietnam, the Russian army against the Chechens and the dictator Saddam Hussein against the Iraqi Kurds. Although white phosphorus inflicts damage similar to chemical weapons, it is still classified as a conventional "incendiary weapon" under international law.

4.4. Missiles

A missile is a self-propelled and guided projectile (unlike a rocket) and is powered by a propellant: either a rocket motor, a booster (commonly a ramjet[15]), or a combination of both (with the rocket providing the initial thrust and later being assisted by a ramjet).

A ballistic missile is defined by its predominantly ballistic trajectory, which is influenced by the *velocity obtained during initial acceleration*, propulsion and the *force of gravity*. The missile has the capacity to carry one or multiple weapons.

Presently, there is an intense race among nations to acquire missiles. One key characteristic is range, as it determines if a country falls within the missile's reach.

Missile defense comprises the tools and strategies used to counter the threat posed by ballistic missiles to military forces in the field and populations at home. Originally developed during the Cold War, missile defense was aimed at protecting American and Soviet territories against intercontinental ballistic missiles. Its development has continued into the 21st century, driven by the proliferation of ballistic missiles in a geopolitical landscape shaped by conflicts in the Middle East and rising tensions in Asia and Europe[16].

4.5. The Space War

The safeguarding of satellite installations has driven the most advanced nations to establish defense structures. France recently formed a specialized defense unit known as the Space Army.

15 See: https://www.futura-sciences.com/sciences/definitions/univers-statoreacteur-2101/.
16 See: https://en.wikipedia.org/wiki/Missile_defense.

We must remain hopeful that this does not exacerbate conflicts between nations. In this context, the importance of peacemaking and comprehensive controlled disarmament is more critical than ever.

4.6. Weapons and pollution

Telescope magazine recently published an intriguing study by Afifa Khazri, a professor at the Royal Military College of Canada, exploring the connection between weapons and pollution[17]. The abstract reads:

> The relationship between social conflicts, the environment and sustainable development is complex, a reality that must be accounted for in any solutions that are contemplated and devised as part of the economic and political decision-making process. In this article, I examine the relationship between sustainable development and armed conflicts. I argue that in order to achieve meaningful sustainable development, international organizations must structure their programs with a view to enabling the poorer nations to emerge from their misery.

There are several interesting websites to consult on this topic[18].

Topics such as toxic substances in ammunition left in the soil after wars is covered, as is the significant carbon footprint of air transport – attributed to its military origins, which prioritized power over efficiency, and the relationship between armed conflicts and the environment (including frameworks, modalities, methods, and the role of environmental assessment). The article includes contributions by Al-Hamandou Dorsouma and Michel-André Bouchard.

An article by Jean-Marc Lavieille in Médiapart (doi: 10.4000/developpementdurable.3365) discusses the extensive environmental damage caused by armed conflicts. In addition, an article by Notre-planète-info examines the severe environmental harm resulting from war, condemning the extensive use of highly toxic substances, the risk of escalation, the manipulation of the environment as a military strategy and the environment as a potential trigger for war.

---

17 Khazri, A. (2011). Le développement durable et les conflits armés. *Télescope*, 17(2), 114–130.

18 See: https://fr.wikipedia.org/wiki/Pollution_induite_par_les_munitions; https://reporterre.net/L-avion-champion-de-la-pollution-Merci-l-armee; https://journals.openedition.org/developpementdurable/3365; https://blogs.mediapart.fr/lavieille/blog/290415/conflits-armes-lampleur-des-atteintes-lenvironnement-i; https://www.notre-planete.info/actualites/3548-guerre_environnement.

4.7. Conclusion

Numerous books, press articles and television shows have depicted the horrors of war.

Therefore, the contents of this chapter will likely not shock readers who are familiar with works such as Leo Tolstoy's *War and Peace* or Francis Ford Coppola's film *Apocalypse Now*.

These technical descriptions aim to provide a realistic understanding of combustion weapons without exaggerating their destructiveness. In addition to flames and chemical reaction bombs, the 20th century witnessed the devastation caused by conventional firearms, weapons of mass destruction (such as atomic bombs, which involve combustion but not in the classical chemical sense) in Japan and bladed weapons forged by fire in Rwanda.

The aftermath of such conflicts underscores the importance of avoiding them, whenever possible. Implementing arms trade controls could play a significant role in this regard, assuming countries agree to adhere to these regulations (see Tony Fortin's article on this subject[19]).

4.8. Appendices

4.8.1. *Appendix 1: reflections on the context and meaning of this chapter*

Gun fire: do we need to justify ourselves before discussing such questions? Are they taboo?

Perhaps it is more of a moral issue than anything else. By highlighting the death and suffering caused by certain weapons, the aim is to encourage societies to avoid these practices and to warn against cruelty and perversity. However, what are the limits of deterrence?

In addressing firearms in this manner, do we risk making not only the combatants but also those who prepare the means of warfare feel guilty? Is that fair? If so, there is more to it than that.

Similar questions can be posed regarding the consequences of policing (a national issue for each country), pollution (international) and social inequality (both national and international).

19 Fortin, T. (2021). Contrôle des ventes d'armes : quel rôle pour les parlementaires ? *Observatoire des armements* [Online]. Available at: http://obsarm.org/spip.php?article353.

These issues are not merely individual concerns; they are societal, political, economic and commercial problems that challenge philosophical thought. It is normal to ask these questions, regardless of our roles in society as citizens or professionals. In this sense, *they are also relevant to scientists* at all levels.

We have a *duty to tell the truth*. Information is essential at every level, as we have seen with the Covid pandemic!

4.8.2. *Appendix 2: additional information*

4.8.2.1. *Classical firearms*

There is extensive documentation on conventional firearms[20].

By definition, the primary purpose of a firearm is to strike a distant target with projectiles (such as balls, bullets or pellets) propelled into the barrel by deflagrating gases from the confined combustion of a chemical mixture. The rate of combustion directly influences the muzzle velocity[21], which in turn determines the effectiveness and range of the projectile. Under normal conditions, *deflagration* limits the velocity of the burnt gases, and therefore the projectile, to about 2,000 m/s.

In modern firearms, projectiles contain their own deflagrating mixture enclosed within the base of a casing or cartridge. The deflagration is initiated by a sudden shock from a firing pin or an electrical discharge.

Firearms – where combustion occurs within a chamber – emerged after the Byzantine conflagration in the 13th century. They were used by the troops of Genghis Khan, then by the Chinese, and later in Europe during the latter part of the Middle Ages. From the 17th century onward, the development of reliable, reusable and cost-effective individual firearms enabled Europe to achieve military dominance.

These weapons became widespread across all countries during the 19th and 20th centuries.

There are two *main categories* of firearms:

1) Light weapons: small-caliber, individual weapons, often less than 15 millimeters, aimed by hand at a visible target.

20 You may find it useful to consult the website (see: https://en.wikipedia.org/wiki/Firearm) from which this paragraph has been extracted in summary form.

21 The muzzle of a gun is the end of the barrel closest to the target.

2) Heavy weapons: larger, collective weapons, such as artillery pieces, which require mounts to be placed in batteries.

Firearms are also classified based on the automation of their firing sequences. They are used in hunting, various sports, crimes and warfare. Throughout history, many types of firearms have emerged, including the harquebus, underwater firearm, fire stick, rifle, shotgun (encompassing assault, hunting, machine, pump action and precision black powder rifles), rocket launchers, machine guns, muskets, pistols, machine pistols, pepper-boxes, revolvers and blunderbuss[22].

The UN considers small arms and light weapons to be *potential or proven sources of exacerbating violence.*

In mid-2007, a report by the Graduate Institute of International Studies in Geneva estimated that 650 million of the world's 850 million small arms (76.5%) are in civilian hands, with the remainder held by various military forces and law enforcement agencies.

The use of certain types of ammunition in wilderness areas, fields and meadows significantly contributes to *environmental pollution*. Each year, tens of thousands of tons of lead are dispersed into the environment through hunting and trap shooting, adding to the cumulative impact of diffuse pollution from previous years. This includes toxic lead from bullets and shots, heavy metals from primers and shooting fumes released into the environment, as well as dispersed toxic waste (DTQD), which poses a risk of lead poisoning to animals, particularly birds and humans.

4.8.2.2. *About napalm*

Napalm inflicts severe burns on exposed individuals due to its flaming projectiles. Its gel-like consistency adheres to the skin and burns tissue down to the bone, with no effective method to halt the burning process. Attempting to cool wounds with water is futile.

Beyond its lethal effects, napalm is notorious for its profound psychological impact on its victims. The "little napalm girl" was just nine years old. As an adult, Kim Phuc harbored resentment toward Nick Ut's photograph, which only depicted her as a survivor of horror in the eyes of the world.

---

22 It is useful to define some of these ancient weapons, for example: the *musket* is a portable firearm with a long, smooth-bore barrel, shoulder stock and lock with a wick or spinning wheel, used by infantry in the 16th and 17th centuries; the *pepper-box* is a type of revolving barrel firearm, a kind of revolver with a very long cylinder, each of the four to six chambers (sometimes eight or more) of which acts directly as a barrel.

In 1980, a United Nations convention and international humanitarian law, grounded in the Geneva Conventions, prohibited the *use of napalm* against civilians. These laws ban all weapons that do not distinguish between civilians and combatants.

4.8.2.3. *On white phosphorus*

Napalm is often combined with white phosphorus to enhance its effect. White phosphorus, which acts as the fuse in this type of incendiary bomb, ignites the napalm. When a napalm bomb detonates on the ground, the phosphorus ignites at a high temperature, leaving a distinctive white trail, and then the napalm ignites. This creates a large, rapidly advancing fireball, which, in the case of the largest incendiary munitions, can reach the size of a multi-story building.

White phosphorus (also known as yellow phosphorus) is a highly flammable white solid that emits corrosive fumes. After the combustion phase, which converts it to phosphorus pentoxide (P_4O_{10}), it absorbs moisture from the air and living tissue to form phosphoric acid ($P_4O_{10} + 6\ H_2O \rightarrow 4\ H_3\ PO_4$). When used in bombs, this results in the dissolution of living tissue (which is primarily composed of water) but does not damage clothing.

White phosphorus can be converted to red phosphorus, an amorphous and harmless allotrope, by heating it to about 250°C. Red phosphorus is the thermodynamically stable allotrope under normal temperature and pressure conditions.

In 1983, the UN classified the offensive use of white phosphorus as a war crime[23] and prohibited it. Israel used white phosphorus as a smoke bomb during the siege of Beirut in 1982, during the offensive in Lebanon in 2006, in Gaza in autumn 2006, and during the 2008–2009 Gaza war with flare shells in urban areas. The Israeli army asserts that their use of these weapons is "within the legal limits of international law". In 2010, Hamas launched phosphorus bombs from Gaza into Israel.

During this period, the International Committee of the Red Cross (ICRC) provided an overview of the rules governing phosphorus weapons and explained their position without confirming Israel's use of white phosphorus in Gaza. However, Human Rights Watch contends that "customary international law" prohibits the use of white phosphorus weapons in densely populated areas like Gaza.

23 See: https://en.wikipedia.org/wiki/White_phosphorus; https://www.francetvinfo.fr/monde/proche-orient/israel-palestine/israel-a-chaque-offensive-de-tsahal-son-arme-polemique_652919.html.

4.8.2.4. *On the Geneva Conventions*

The Geneva Conventions are key international treaties in the realm of international humanitarian law. They outline the rules of conduct to be adhered to during armed conflict, particularly regarding the protection of civilians, humanitarian aid workers, the wounded and prisoners of war.

The first Geneva Convention dates back to 1864. However, the current texts were drafted after World War II. There are seven documents currently in force: the four Geneva Conventions of August 12, 1949, the two Additional Protocols of June 8, 1977 and the Third Additional Protocol of 2005. The four Geneva Conventions have been ratified globally, meaning every country in the world is obligated to uphold them.

5

Shifting Interface in Spacetime

The macroscopic balance equations of the fluid interfaces were established by considering two scales of length and by means of some approximations, such as the conservation of the mixed velocity vector and of the gradient parallel to the crossing of the interfacial zone. We generalize this method to the case of four interfaces in the presence of electromagnetic fields. The transition to spacetime is indeed a means of obtaining a homogeneous presentation of the balance sheets.

List of symbols

Latin characters

A, A^+, AS	chemical species, positive ion of A, adsorbed species
$\vec{\mathbf{B}}$	magnetic field vector
c	speed of light in vacuum in the absence of a field
d	distance; differential
dl	elementary distance
ds	elementary interval between events
$\vec{\mathbf{D}}$	electric displacement vector

Chapter written by Roger PRUD'HOMME.

$_4\underline{\underline{\mathbf{D}}}$	matrix of the special Lorentz transformation
e	free electron; internal mass energy
E	internal energy (per unit mass)
$\vec{\mathbf{E}}$	electric field vector
F	physical quantity (f per unit of mass)
$\vec{f}$	force acting at a distance on the total unit of mass
$\vec{f}_k$	force acting at a distance on the unit of mass of species
G	free enthalpy (per unit mass, for a species in a mixture)
$\vec{\mathbf{I}}$	classical current vector $\tilde{\rho}\,\underline{\mathbf{v}}$; total current $\tilde{\rho}\,\underline{\mathbf{v}} + \underline{\mathbf{i}}$
$_4\boldsymbol{I}$	4-current vector
i	diffusion current intensity
$\vec{\mathbf{i}}$	total conduction (or diffusion) current vector
I	electric current intensity equal to $\lvert\vec{\mathbf{I}}\rvert$
$\mathbf{J}_{VF}$	flow of a quantity F relative to the speed movement $\vec{\mathbf{V}}$
$\mathbf{J}_F$	flow of F relative to a fixed reference frame $\vec{\mathbf{V}} = \vec{\mathbf{0}}$
$\mathcal{J}_F$	flux of F relative to the barycentric motion of the fluid $\vec{\mathbf{V}} = \vec{\mathbf{v}}$
$\vec{\mathcal{J}}_{Dk}$	diffusion flux of the species k equal to $\rho_k\left(\mathbf{v}_k - \mathbf{v}\right)$
k	chemical species; Boltzmann constant
K	kinetic energy (k per unit mass)

m	total mass; mass of a material point m_0/α
m_0	mass of a material point in Minkowski space
M	mass; molecular mass; material point
$\mathcal{M}$:	molar mass
$\mathbf{M}$	magnetic polarization ($\mathbf{m} = \mathbf{M}/\rho$: per unit mass)
$_4\overrightarrow{\mathbf{M}}$	4-vector corresponding to the material point
$_4\boldsymbol{\mathcal{M}}$	4-tensor equal to $_4\boldsymbol{\mathcal{F}} - {}_4\boldsymbol{\mathcal{H}}$
$\dot{m}$	mass flow rate per unit area; unit flow rate
n	total number of moles
n_k	number of moles of species k
N	number of species; normal coordinate of an interface
$\vec{\mathbf{n}}, \vec{\mathbf{N}}, {}_4\overrightarrow{\mathbf{N}}$	unit normal to a surface, 4-normal to a surface in spacetime
p	thermodynamic pressure
$\vec{\mathbf{P}}$	electric polarization vector ($\mathbf{p} = \vec{\mathbf{P}}/\rho$; per unit mass)
$\vec{\vec{\mathbf{P}}}$	pressure tensor
$_4\boldsymbol{\mathcal{P}}$	4-momentum-energy tensor
$_4\boldsymbol{\mathcal{P}}^{ch}$	4-momentum-energy tensor due to electromagnetic field (or 4-electromagnetic flux tensor) $_4\boldsymbol{\mathcal{P}}^{(m)}$: 4-momentum-energy tensor due to mass
$\vec{\mathbf{q}}$	heat flux vector
R	universal molar gas constant; radius; or electrical resistance

$\Re$	reference
S	entropy (s per unit mass)
S, $\mathcal{S}$	surface
t	time
T	absolute temperature
u, v, w	velocity components $\vec{\mathbf{v}}$ in Cartesian coordinates
$\vec{\mathbf{v}}$	velocity vector; velocity vector of the material coordinate point v_x, v_y and v_z of module v; barycentric velocity vector in a composite fluid $\sum_k Y_k \vec{\mathbf{v}}_k$
$\vec{\mathbf{v}}_k$	velocity vector of the species k
$\vec{\mathbf{V}}$	vector; speed vector; mixed speed at the interface level $\vec{\mathbf{V}} = \vec{\mathbf{v}}_{//} + \vec{\mathbf{w}}$
$_4\mathbf{V}$	4-material point velocity
$\mathcal{V}$, $(\partial \mathcal{V})$	control volume, its boundary
$\vec{\mathbf{V}}_k$	speed of diffusion of species $\vec{\mathbf{v}}_k - \vec{\mathbf{v}}$
$\vec{\mathbf{w}}$	speed of a surface (its normal component)
$\dot{W}_F, \dot{W}_{Fa}$	rate of production of the quantity F resp. per unit of volume, per unit of area
$\dot{W}_k$	species production rate k
$\dot{W}_{E_t a}$	total energy source per unit area and time at the interface
x, y, z	Cartesian coordinates

X_k, Y_k	molar fraction, resp. mass fraction, of the species k
z	mass charge $\sum_k z_k Y_k$ of a mixture
z_k	species mass charge k

Greek characters

α	quantity $\sqrt{1 - v^2/c^2}$
$\underline{\underline{\boldsymbol{\varepsilon}}}$	dielectric tensor
$\vec{\varphi}, \underline{\varphi}, {}_4\underline{\varphi}$	vector, column matrix of the vector, column matrix of the corresponding four-vector
$\vec{\vec{\mathbf{T}}}, \underline{\underline{\mathbf{T}}}, , {}_4\underline{\underline{\mathbf{T}}}$	tensor of order 2, matrix of this tensor, matrix of a four-tensor
${}_4\vec{\mathbf{\Phi}}$	4-force vector
$\vec{\vec{\Pi}}$	viscous pressure tensor
ρ	density
ρ	partial density of the species j
τ	proper time of a material point in Minkowski space equal to αt
$\tilde{\rho}$:	classic charge per unit volume $\tilde{\rho}_0/\alpha$
$\tilde{\rho}_0$	electric charge in the Minkowski space
$\tilde{\rho}_k$	charge of the species k per unit volume
ϑ	volume per unit mass equal to $1/\rho$

Other symbols, subscripts, superscripts

a	subscript designating an extensive quantity per unit area
ch	superscript designating a *field-type* quantity
m	superscript designating a *mass-type* quantity
s	subscript designating an intensive or specific quantity at the interface
T	superscript indicating the transposition of a tensor, a matrix
$\vec{\nabla}$	gradient (nabla)
$//$, $\perp$	parallel, resp. perpendicular to a surface
$\vec{\vec{1}}$	unit tensor

5.1. Introduction

Let us summarize the objectives of this chapter. The issues addressed are as follows:

1) Correctly write the equations at the interfaces in the 4D Minkowski space.

2) Write these equations in the presence of an electromagnetic field.

The interest of a 4D formulation is to: obtain tensor formulations and consider, if necessary, relativity.

– *The method adopted consists* of three points: we recall the method used in the 3D case for fluids in the absence of a field, first in the volume, then at the interface.

– The rules of special relativity are then recalled as being applied to the balances in the four-volume, then to the interface.

– The electromagnetic tensors are then defined, which makes it possible to establish the balances in fluid media in the presence of fields.

5.2. Reminder concerning interfaces in 3D coordinates

5.2.1. *Volume balances*

It is interesting to write *the balance equations in the volume* by following the motion having a known velocity field. Since F is an extensive physical property (mass, momentum, energy, etc.) of a fluid (more generally, of a continuous medium), of value f per unit of mass, we can write *the integral form of the balance of F in the volume*:

$$\frac{d}{dt}\int_{\mathcal{V}} \rho f \, d\mathcal{V} + \int_{\partial\mathcal{V}} \vec{\jmath}_F \cdot \vec{n} \, dS = \int_{\mathcal{V}} \dot{W}_F \, d\mathcal{V}$$

$(\mathcal{V})$: limit surface control volume $(\partial\mathcal{V})$ having as external normal $\vec{n}$ (Figure 5.1).

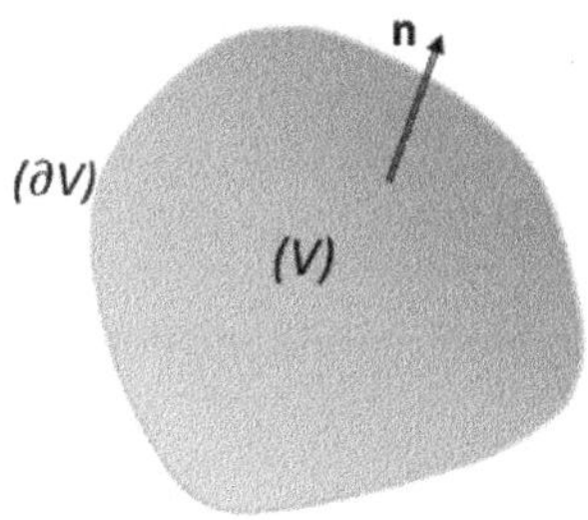

2

Figure 5.1. *Connected control volume and unit normal to its boundary For any quantity F (f per unit of mass), we will therefore write locally:*

$$\frac{d_{\mathbf{V}}(\rho f)}{dt} + \rho f \vec{\nabla} \cdot \vec{\mathbf{V}} + \vec{\nabla} \cdot \vec{\mathbf{J}}_{\mathbf{V}F} = \dot{W}_F \qquad [5.1]$$

with the following definitions of the derivative by following the speed movement $\vec{\mathbf{V}}$, and of the flux vector: $\frac{d_{\mathbf{V}}}{dt} = \frac{\partial}{\partial t} + \vec{\mathbf{V}} \cdot \vec{\nabla}$, $\vec{\mathbf{J}}_{\mathbf{V}F} = \vec{\jmath}_F + \rho f \left(\vec{\mathrm{v}} - \vec{\mathbf{V}}\right)$, the flux $\vec{\jmath}_F$ being that of the barycentric motion (speed $\vec{\mathrm{v}}$). The rate of production of the unit volume of the quantity F is noted $\dot{W}_F$.

We directly deduce from equation [5.1], for example, the mass balance of a simple fluid:

$$\frac{d_{\mathbf{V}}\rho}{dt}+\rho\vec{\nabla}\cdot\vec{\mathbf{V}}+\vec{\nabla}\cdot\rho\left(\vec{\mathbf{v}}-\vec{\mathbf{V}}\right)=0$$

Taking $\vec{\mathbf{V}}=\vec{\mathbf{v}}$: $\frac{d\rho}{dt}+\rho\vec{\nabla}\cdot\vec{\mathbf{v}}=0$, and taking $\vec{\mathbf{V}}=\vec{\mathbf{0}}$: $\frac{\partial\rho}{\partial t}+\vec{\nabla}\cdot(\rho\vec{\mathbf{v}})=0$. These relationships remain valid for a mixture of total density ρ and barycentric velocity $\vec{\mathbf{v}}$. The balances of momentum and energy will also verify equations of the form [5.1], as will the balance of entropy. These are "mass-type quantities". Let us add the balance of electric charge for conductive media.

The case of quantities characterizing the electromagnetic field treated in sections 5.3.3 and 5.3.4 requires specific treatment. The balance equations of these "field-type quantities" will depend in particular on the polarization of the medium (Prud'homme 2014).

5.2.2. Balances at the interfaces

For the interfacial zones, a priori mobile and deformable, it is necessary to define a ***mixed velocity field*** $\vec{\mathbf{V}}$.

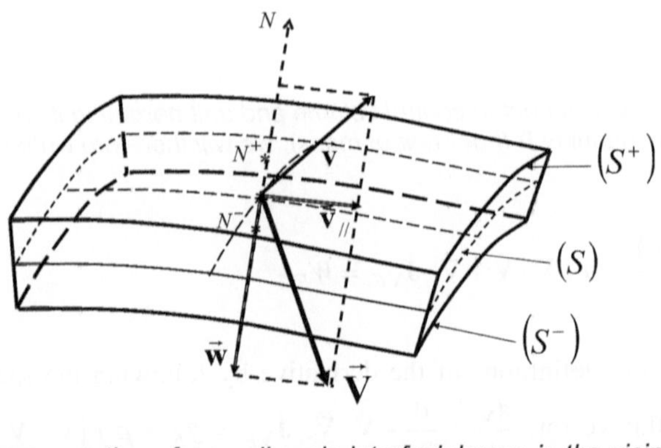

Figure 5.2. *Representation of a small-scale interfacial zone in the vicinity of a current point (*) of the mean surface (S), of normal N oriented from (-) to (+). We define the mixed speed* $\vec{\mathbf{V}}=\vec{\mathbf{v}}_{//}+\vec{\mathbf{w}}$*, where* $\vec{\mathbf{v}}_{//}$ *is the component of the barycentric velocity vector* $\vec{V}$ *fluid parallel to the interface, and* $\vec{\mathbf{w}}$ *is the normal speed of movement of the interface. N denotes the ordinate following the normal N. The ordinates at the current point of (S) and of the boundary surfaces (S⁻) and (S⁺), are, respectively,* $N = N^* = 0$, $N = N^-$ *and* $N = N^+$

The mixed velocity considered when crossing the interfacial zone, shown schematically in Figure 5.2, limited by surfaces S^- and S^+, is $\vec{\mathbf{V}} = \vec{\mathbf{v}}_{//} + \vec{\mathbf{w}}$. We will make specific hypotheses of conservation at the crossing between the normal coordinates N^- and N^+.

In this definition, two movements of different natures intervene: on the one hand, the component parallel to the interface of the barycentric velocity vector of the fluid and, on the other hand, the normal speed of displacement $\vec{\mathbf{w}}$ of the interface (Gatignol and Prud'homme 2001), which explains the name mixed speed.

The *local interface balance* is obtained by integration between N^- and N^+ of equation [5.1] written in the form: $\dfrac{d_{\mathbf{V}}(\rho f)}{dt} + \rho f \vec{\nabla}_{//} \cdot \vec{\mathbf{V}} + \vec{\nabla}_{//} \cdot \vec{\mathbf{J}}_{\mathbf{V}F} + \dfrac{\partial \vec{\mathbf{J}}_{\mathbf{V}F}}{\partial N} = \dot{W}_F$, assuming ***conservation of the mixed velocity vector*** $\vec{\mathbf{V}}$ and the ***conservation of the parallel gradient*** $\vec{\nabla}_{//}$. The expressions of $d_{\mathbf{V}}/dt$ and of $\vec{\mathbf{J}}_{\mathbf{V}F}$ depend on this velocity field $\vec{\mathbf{V}}(x,y,z,t)$.

By posing:

$$\rho_a = \int_{N^-}^{N^+} \rho \, dN \, , \; \rho_a f_S = \int_{N^-}^{N^+} \rho f \, dN , \; \vec{\mathbf{J}}_{\mathbf{V}Fa} = \int_{N^-}^{N^+} \vec{\mathbf{J}}_{\mathbf{V}F} \, dN \;\; , \; \dot{W}_{Fa} = \int_{N^-}^{N^+} \dot{W}_F \, dN \, ,$$

we obtain:

$$\frac{d_S(\rho_a f_S)}{dt} + \rho_a f_S \vec{\nabla}_S \cdot \vec{\mathbf{V}}_S + \vec{\nabla}_S \cdot \vec{\mathbf{J}}_{\mathbf{V}Fa} + \left[J_{\mathbf{V}F\perp}\right]_-^+ = \dot{W}_{Fa} \qquad [5.2]$$

with $\vec{\mathbf{V}} = \vec{\mathbf{V}}_S$, $\vec{\nabla}_{//} = \vec{\nabla}_S$, $d_S/dt = \partial/\partial t + \vec{\mathbf{V}}_S \cdot \vec{\nabla}$, $\phi_\perp = \vec{\phi} \cdot \vec{\mathbf{N}}$, $[\phi]_-^+ = \phi_+ - \phi_-$, $\vec{\mathbf{N}}$ being the unit normal to the interface.

We must also write consistency conditions indicating that the normal components of the interface tensors are zero. Thus, $\vec{\jmath}_{\mathbf{F}a} \cdot \vec{\mathbf{N}} = 0$.

5.3. Minkowski volume balance equations in spacetime

5.3.1. *General (Prud'homme 2014) (special relativity*[1]*, spacetime*[2]*)*

In terms of relativity, we will limit ourselves to special relativity and Galilean frames of reference with low speeds V, such as $V/c \ll 1$. The use of spacetime coordinates and the Lorentz transformation is essential. The Lorentz transformation preserves intervals (whereas in classical mechanics, changes in inertial frames of reference preserve distances). This transformation, presented in Appendices 1 and 2 (sections 5.6.1 and 5.6.2), is in fact a rotation in spacetime. When introducing a tensor physical quantity, it is necessary to ensure that its expression obeys the Lorentz transformation during changes in inertial frames of reference.

Figure 5.3. *Albert Einstein (1879–1955)*

First, we will express the 3D balances of section 5.2 in spacetime. Recall that the results of this section were established in the absence of an electromagnetic field. Recall that these are material quantities of the mass type. The 4D balance equations obtained will therefore be equivalent to the classical 3D equations.

Second, we will deal with the media in the presence of an electromagnetic field. We will then introduce electromagnetic four-tensors, and we will have to express

1 *Einstein's principle of relativity*: the speed of propagation of interactions is constant and equal to the speed of light in a vacuum c. In relativistic mechanics, these are the intervals that are preserved during changes in inertial frames of reference.

2 In space-time, the elementary *interval* between events is defined by: $ds^2 = c^2dt^2 - dl^2$, where dl is the elementary distance such that:$dl^2 = dx^2 + dy^2 + dz^2$.

Maxwell's equations in the Minkowski space. We will then be able to express 4D balances in the presence of an electromagnetic field.

5.3.2. *4D balances in the volume for mass-type quantities*

The speed four-vector $4\vec{\mathbf{V}}$ is defined by the column matrix: $\underline{4\mathbf{V}} = \begin{bmatrix} \vec{\mathbf{V}}/\alpha \\ ic/\alpha \end{bmatrix}$, $\alpha = \sqrt{1 - \frac{|\mathbf{V}|^2}{c^2}}$. Similarly, we define the four-gradient[3] as follows: $\underline{4\nabla} = \begin{bmatrix} \underline{\nabla} \\ (1/ic)(\partial/\partial t) \end{bmatrix}$.

The local four-volume balance sheet is then written as:

$$4\vec{\nabla}_{\mathbf{V}} \cdot 4\vec{\mathbf{J}}_{\mathbf{V}F} + \rho f \, 4\vec{\nabla}_{\mathbf{V}} \cdot 4\vec{\mathbf{V}} = 4\dot{W}_F \qquad [5.3]$$

and the local quad interface balance becomes:

$$4\vec{\nabla}_S \cdot 4\vec{\mathbf{J}}_{VFa} + \rho_a f_S \, 4\vec{\nabla}_S \cdot 4\vec{\mathbf{V}}_S + \left[4\mathbf{J}_{VF\perp}\right]_-^+ = 4\dot{W}_{Fa} \qquad [5.4]$$

with: $\underline{4\mathbf{J}_{\mathbf{V}F}} = \begin{bmatrix} \underline{\mathbf{J}_{\mathbf{V}F}}/c \\ i\rho f \end{bmatrix}$, $\underline{4\mathbf{J}_{\mathbf{V}Fa}} = \begin{bmatrix} \underline{\mathbf{J}_{\mathbf{V}Fa}}/c \\ i\rho_a f_S \end{bmatrix}$, $\underline{4\nabla_S} = \begin{bmatrix} \underline{\nabla_S} \\ (1/ic)\, d_S/dt \end{bmatrix}$, $4\mathbf{J}_{\mathbf{V}F\perp} = 4\vec{\mathbf{J}}_{\mathbf{V}F} \cdot 4\vec{\mathbf{N}}$, $\underline{4\mathbf{N}} = \begin{bmatrix} \underline{\mathbf{N}} \\ i\vec{\mathbf{V}}_S \cdot \vec{\mathbf{N}}/c \end{bmatrix}$ and $4W_F = W_F / c$, respectively, for four-streams, four-gradients and four-productions.

It is necessary to add consistency conditions such as: $4\vec{\mathcal{J}}_{\mathbf{F}a} \cdot 4\vec{\mathbf{N}} = 0$.

---

3 The Laplacian $\Delta = \nabla^2 = \frac{\partial^2}{\partial x^2} + \frac{\partial^2}{\partial y^2} + \frac{\partial^2}{\partial z^2}$ becomes in 4D the d'Alembert $\square = -{}_4\nabla^2 = -\Delta + \frac{1}{c^2}\frac{\partial^2}{\partial t^2}$.

5.3.3. *Electromagnetic quad tensors in the volume*

Electromagnetic fields and movements

We first introduce the following three four-tensors, containing the components of the electromagnetic field and displacement vectors: $\vec{\mathbf{E}},\vec{\mathbf{D}}$ for the electric field part and $\vec{\mathbf{B}},\vec{\mathbf{H}}$ for the magnetic part. The matrices corresponding to these quad tensors are written in the given frame of reference:

$$\underline{\underline{4\mathcal{F}}} = \begin{bmatrix} 0 & B_z & -B_y & -iE_x \\ -B_z & 0 & B_x & -iE_y \\ B_y & -B_x & 0 & -iE_z \\ iE_x & iE_y & iE_z & 0 \end{bmatrix}, \quad \underline{\underline{4\mathcal{F}^\star}} = \begin{bmatrix} 0 & -iE_z & iE_y & B_x \\ iE_z & 0 & -iE_x & B_y \\ -iE_y & iE_x & 0 & B_z \\ -B_x & -B_y & -B_z & 0 \end{bmatrix}$$

$$\underline{\underline{4\mathcal{H}}} = \begin{bmatrix} 0 & H_z & -H_y & -iD_x \\ -H_z & 0 & H_x & -iD_y \\ H_y & -H_x & 0 & -iD_z \\ iD_x & iD_y & iD_z & 0 \end{bmatrix}$$

Figure 5.4. *James Clerk Maxwell (1831–1879)*

Electromagnetic tensors $4\vec{\vec{\mathcal{H}}}$ and $4\vec{\vec{\mathcal{F}}}^\star$ verify *the 4D expressions of Maxwell's equations*:

$$4\vec{\nabla} \cdot 4\vec{\vec{\mathcal{H}}} = 4\vec{I}\ ,\ 4\vec{\nabla} \cdot 4\vec{\vec{\mathcal{F}}}^\star = \vec{0} \qquad [5.5]$$

We further consider the momentum-energy four-tensor: $4\vec{\vec{\mathcal{P}}} = 4\vec{\vec{\mathcal{P}}}^m + 4\vec{\vec{\mathcal{P}}}^{ch}$, *with its part due to mass* and *its part due to fields*. These four-tensors represent the fluxes of momentum-energy. $4\vec{\vec{\mathcal{P}}}^{ch}$ is a function of the electric and magnetic displacement fields: $\vec{\mathbf{E}},\vec{\mathbf{B}},\vec{\mathbf{D}},\vec{\mathbf{H}}$ and the velocity vector $\vec{\mathbf{v}}$ reported to the speed c of light in vacuum.

We define the following tensors and corresponding matrices:

– *four-tensor of momentum energy due to mass*, **in volume**:

$$\underline{\underline{4\mathcal{P}}}^{m} = \begin{bmatrix} \rho\vec{\mathbf{v}}\otimes\vec{\mathbf{v}}+\vec{\vec{\mathbf{P}}} & ic\rho\vec{\mathbf{v}} \\ \frac{i}{c}\left[\vec{\mathbf{q}}+\vec{\vec{\mathbf{P}}}\cdot\vec{\mathbf{v}}+\rho(e+k)\vec{\mathbf{v}}\right] & -\rho(e+k) \end{bmatrix}$$

– *four-tensor of pressure energy due to mass*, **at the interface**:

$$\underline{\underline{4\mathcal{P}}}_{a}^{\ m} = \begin{bmatrix} \rho_a\vec{\mathbf{v}}_S\otimes\vec{\mathbf{v}}_S+\vec{\vec{\mathbf{P}}}_a & ic\rho_a\vec{\mathbf{v}}_S \\ \frac{i}{c}\left[\vec{\mathbf{q}}_a+\vec{\vec{\mathbf{P}}}_a\cdot\vec{\mathbf{v}}_S+\rho_a(e_S+k_S)\vec{\mathbf{v}}_S\right] & -\rho_a(e_S+k_S) \end{bmatrix}$$

– *four-tensor of momentum-energy due to the field*:

$$\underline{\underline{4\mathcal{P}}}^{ch} = \begin{bmatrix} \underline{\underline{\mathbf{P}}}^{ch} & i\underline{\mathbf{E}}\times\underline{\mathbf{H}} \\ (i/c)\,\underline{\mathbf{J}_E^{ch}} & -E^{ch} \end{bmatrix}$$

For fluid moving at barycentric velocity $\vec{\mathbf{v}}$:

$$\vec{\vec{\mathbf{P}}}^{ch} = \left(\frac{\left|\vec{\mathbf{E}}\right|^2+\left|\vec{\mathbf{B}}\right|^2}{2}-\vec{\mathbf{M}}\cdot\vec{\mathbf{B}}\right)\vec{\vec{\mathbf{1}}}-\vec{\mathbf{D}}\otimes\vec{\mathbf{E}}-\vec{\mathbf{B}}\otimes\vec{\mathbf{H}}-\frac{\vec{\mathbf{v}}}{c}\otimes\left(\vec{\mathbf{P}}\times\vec{\mathbf{B}}-\vec{\mathbf{M}}\times\vec{\mathbf{E}}\right)$$

$$\vec{\mathbf{J}}_E^{ch} = c\left(\vec{\mathbf{E}}\times\vec{\mathbf{H}}\right)-\left(\vec{\mathbf{P}}\cdot\vec{\mathbf{E}}+\vec{\mathbf{M}}\cdot\vec{\mathbf{B}}\right)\vec{\mathbf{v}},$$

The vector product $\vec{\mathbf{E}}\times\vec{\mathbf{H}}$ is called the “Poynting vector”.

$$E^{ch} = \frac{\left|\vec{\mathbf{E}}\right|^2+\left|\vec{\mathbf{B}}\right|^2}{2}-\vec{\mathbf{M}}\times\vec{\mathbf{B}}-\frac{\vec{\mathbf{v}}}{c}\cdot\left(\vec{\mathbf{P}}\times\vec{\mathbf{B}}-\vec{\mathbf{M}}\times\vec{\mathbf{E}}\right)$$

The fields and displacements being here measured in the reference frame at rest (according to de Groot and Mazur 1984) and in the non-relativistic case.

– Four-vector of external forces: $\underline{4\boldsymbol{\Phi}}_k = \begin{bmatrix} \rho_k \vec{f}_k \\ \dfrac{i}{c}\rho_k \vec{f}_k \cdot \vec{\mathbf{v}}_k \end{bmatrix}$

$$_4\vec{\boldsymbol{\Phi}} = \sum_k {}_4\vec{\boldsymbol{\Phi}}_k$$

5.3.4. *4D momentum-energy balance in the volume*

We write that the outgoing momentum-energy flux across the boundary of the four-volume considered is equal to the four-vector of the external forces, which locally gives:

$$_4\vec{\nabla} \cdot \left({}_4\vec{\vec{\mathcal{P}}}^{ch} + {}_4\vec{\vec{\mathcal{P}}}^{m} \right) = {}_4\vec{\boldsymbol{\Phi}} \qquad [5.6]$$

Figure 5.5. *Lev Davidovitch Landau (1908–1968)*

5.4. The balance equations at the interface in the Minkowski spacetime

Let us use interfaces with electromagnetic field, the same organization as for the volume in section 5.3: knowing the interface balances in the absence of electromagnetic field, we define the interface electromagnetic quantities and the momentum-energy tensors. Then, we are interested in Maxwell's equations and general balances by limiting ourselves to the case of simple fluids.

5.4.1. *Electromagnetic four-tensors*

We give in the following the basic definitions and equations:

– *Electromagnetic fields and displacements*: as in the case of volume, we can define interface quantities, which must verify Maxwell's equations and coherence conditions:

$$\underline{\underline{4\mathcal{F}}}_a = \begin{bmatrix} 0 & B_{az} & -B_{ay} & -iE_{ax} \\ -B_{az} & 0 & B_{ax} & -iE_{ay} \\ B_{ay} & -B_{ax} & 0 & -iE_{az} \\ iE_{ax} & iE_{ay} & iE_{az} & 0 \end{bmatrix},$$

$$\underline{\underline{4\mathcal{F}}}_a^{\,\star} = \begin{bmatrix} 0 & -iE_{az} & iE_{ay} & B_{ax} \\ iE_{az} & 0 & -iE_{ax} & B_{ay} \\ -iE_{Sy} & iE_{ax} & 0 & B_{az} \\ -B_{ax} & -B_{ay} & -B_{az} & 0 \end{bmatrix}$$

$$\underline{\underline{4\mathcal{H}}}_a = \begin{bmatrix} 0 & H_{az} & -H_{ay} & -iD_{ax} \\ -H_{az} & 0 & H_{ax} & -iD_{ay} \\ H_{ay} & -H_{ax} & 0 & -iD_{az} \\ iD_{ax} & iD_{ay} & iD_{az} & 0 \end{bmatrix}$$

The interface quantities are in a *polarized medium*: $\mathbf{H}_a$ is the magnetic displacement vector; $\mathbf{D}_a$ is the electric displacement vector; $\mathbf{B}_a$ is the magnetic field; $\mathbf{E}_a$ is the electric field; $\vec{\mathbf{P}}_a = \mathbf{D}_a - \mathbf{E}_a$ is the electric polarization ($\mathbf{p}_S$ per unit mass); $\mathbf{M}_a = \mathbf{B}_a - \mathbf{H}_a$ is the magnetic polarization ($\mathbf{m}_S$ per unit mass).

– *Maxwell's equations at the interface and coherence conditions*:

$$4\vec{\nabla} \cdot 4\vec{\vec{\mathcal{H}}}_a = 4\vec{I}_a \;,\; 4\vec{\nabla} \cdot 4\vec{\vec{\mathcal{F}}}_a^{\,\star} = \vec{\mathbf{0}} \qquad [5.7]$$

Conditions of consistency:

$$\boxed{{}_4\vec{\vec{\mathcal{H}}}_a \cdot {}_4\vec{\mathbf{N}} = {}_4\vec{\vec{\mathcal{F}}}*_a \cdot {}_4\vec{\mathbf{N}} = \vec{\mathbf{0}}, \quad {}_4\vec{\mathbf{I}}_a \cdot {}_4\vec{\mathbf{N}} = 0}$$

– *Momentum-energy four-tensors at the interface and coherence conditions*:

$$\underline{\underline{\mathcal{P}}}_a{}^m = \begin{bmatrix} \rho_a \vec{\mathbf{v}}_S \otimes \vec{\mathbf{v}}_S + \vec{\vec{\mathbf{P}}}_a & ic\rho_a \vec{\mathbf{v}}_S \\ \dfrac{i}{c}\left[\vec{\mathbf{q}}_a + \vec{\vec{\mathbf{P}}}_a \cdot \vec{\mathbf{v}}_S + \rho_a (e_S + k_S)\vec{\mathbf{v}}_S\right] & -\rho_a (e_S + k_S) \end{bmatrix}$$

$$\underline{\underline{4\mathcal{P}}}_a{}^{ch} = \begin{bmatrix} \vec{\vec{\mathbf{P}}}_a{}^{ch} & ic\left(\rho_a \vec{\mathbf{v}}_S + \dfrac{1}{c}\vec{\mathbf{E}}_a \times \vec{\mathbf{H}}_a\right) \\ (i/c)\vec{\mathbf{J}}_{Ea}{}^{ch} & -E_a{}^{ch} \end{bmatrix}$$

Consistency conditions:

$${}_4\vec{\vec{\mathcal{P}}}_a{}^m \cdot {}_4\vec{\mathbf{N}} = \vec{\mathbf{0}}, \quad {}_4\vec{\vec{\mathcal{P}}}_a{}^{ch} \cdot {}_4\vec{\mathbf{N}} = \vec{\mathbf{0}}, \ {}_4\vec{\mathcal{J}}_{Fa} \cdot {}_4\vec{\mathbf{N}} = 0, \quad {}_4\vec{\mathbf{J}}_{Fa} \cdot {}_4\vec{\mathbf{N}} = 0,$$

$$4\vec{\vec{\mathcal{H}}}_a \cdot 4\vec{\mathbf{N}} = \vec{\mathbf{0}}, \quad 4\vec{\vec{\mathcal{F}}}_a * \cdot \vec{\mathbf{N}} = \vec{\mathbf{0}} \text{ and } 4\vec{\vec{\mathcal{P}}}^m \cdot 4\vec{\mathbf{N}} = \vec{\mathbf{0}}, \ 4\vec{\vec{\mathcal{P}}}^{ch} \cdot 4\vec{\mathbf{N}} = \vec{\mathbf{0}}$$

or, expanding into three-dimensional space:

$$\begin{cases} \left(\rho_a \vec{\mathbf{v}}_S \otimes \vec{\mathbf{v}}_S + \vec{\vec{\mathbf{P}}}_a\right) \cdot \vec{\mathbf{N}} - V_{S\perp} \rho_a \vec{\mathbf{v}}_S = \vec{\mathbf{0}} \\ \left[\vec{\mathbf{q}}_a + \vec{\mathbf{v}}_S \cdot \vec{\vec{\mathbf{P}}}_a + \rho_a (e_a + k_a)\vec{\mathbf{v}}_S\right]_-^+ \cdot \vec{\mathbf{N}} - \rho_a (e_a + k_a) V_{S\perp} = 0 \\ \vec{\vec{\mathbf{P}}}_a{}^{ch} \cdot \vec{\mathbf{N}} - \dfrac{V_{S\perp}}{c} \vec{\mathbf{E}}_a \times \vec{\mathbf{H}}_a = \vec{\mathbf{0}} \\ \vec{\mathbf{J}}_{Ea}{}^{ch} \cdot \vec{\mathbf{N}} - V_{S\perp} E_a{}^{ch} = 0 \end{cases}$$

$$\begin{cases} \vec{\vec{\mathbf{P}}}_a \cdot \vec{\mathbf{N}} = \mathbf{0}, \ \vec{\vec{\mathbf{P}}}_a{}^{ch} \cdot \vec{\mathbf{N}} = \mathbf{0} \\ \vec{\mathbf{q}}_a \cdot \vec{\mathbf{N}} = 0, \ \vec{\mathbf{E}}_a{}^2 - \vec{\mathbf{B}}_a{}^2 = 0 \end{cases}$$

5.4.2. *Momentum-energy balance at the interface*

We give below the balance equations, in which the electromagnetic quantities are globalized for a mixture, as is also the pressure tensor.

– *4D pulse-energy balance at the interface*:

$$\boxed{\begin{aligned}&{}_4\vec{\nabla}_S\cdot\left({}_4\vec{\vec{\mathcal{P}}}_a^{ch}+{}_4\vec{\vec{\mathcal{P}}}_a^{m}\right)+\rho_a\vec{\mathbf{v}}_S\,{}_4\vec{\nabla}_S\cdot{}_4\vec{\mathbf{V}}_S+\\&\left[{}_4\vec{\vec{\mathcal{P}}}^{ch}+{}_4\vec{\vec{\mathcal{P}}}^{m}+(1/c)\rho\left(\vec{\mathbf{v}}-\vec{\mathbf{v}}_S\right)\otimes\left(\vec{\mathbf{v}}-\vec{\mathbf{V}}\right)\right]_-^+\cdot\vec{\mathbf{N}}={}_4\vec{\Phi}_a\end{aligned}} \quad [5.8]$$

or, in classical writing, k designates the chemical species:

$$\rho_a\frac{d_S\vec{\mathbf{v}}_S}{dt}+\vec{\nabla}_S\cdot\vec{\vec{\mathbf{P}}}_a+\left[\vec{\vec{\mathbf{P}}}+\rho\left(\vec{\mathbf{v}}-\vec{\mathbf{v}}_S\right)+\vec{\vec{\mathbf{P}}}^{ch}-\frac{1}{c}\left(\vec{\mathbf{E}}\times\vec{\mathbf{B}}\right)\otimes\vec{\mathbf{V}}\right]_-^+\cdot\vec{\mathbf{N}}=\sum_k\rho_{ak}\vec{f}_{ka}$$

$$\rho_a\frac{d_Se_S}{dt}+\vec{\nabla}_S\cdot\vec{\mathbf{q}}_a+\left[\vec{\mathbf{q}}+\rho\left(e-e_S\right)\left(\vec{\mathbf{v}}-\vec{\mathbf{V}}\right)\right]_-^+\cdot\vec{\mathbf{N}}=-\vec{\vec{\mathbf{P}}}_a:\vec{\nabla}_S\otimes\vec{\mathbf{v}}_S+\sum_k\vec{\jmath}_{Dka}\cdot\vec{f}_{ka}$$

$$-\left[\left(\vec{\mathbf{v}}-\vec{\mathbf{v}}_S\right)\cdot\vec{\vec{\mathbf{P}}}+\frac{1}{2}\left(\vec{\mathbf{v}}-\vec{\mathbf{v}}_S\right)^2\left(\vec{\mathbf{v}}-\vec{\mathbf{V}}\right)\right]_-^+\cdot\vec{\mathbf{N}}+\dot{W}_{Ea}{}^{ch}$$

– *Species and total mass*:

$$\frac{d_S\left(\rho_a Y_{kS}\right)}{dt}+\rho_a Y_{kS}\vec{\nabla}_S\cdot\vec{\mathbf{V}}_S+\vec{\nabla}_S\cdot\vec{\mathbf{J}}_{\mathbf{V}ka}+\left[\vec{\jmath}_{Dk}+\rho Y_k\left(\vec{\mathbf{v}}-\vec{\mathbf{V}}\right)\right]_-^+\cdot\vec{\mathbf{N}}=\dot{W}_{ka} \quad [5.9]$$

$$\frac{d_S\rho_a}{dt}+\rho_a\vec{\nabla}_S\cdot\vec{\mathbf{V}}_S+\left[\rho\left(\vec{\mathbf{v}}-\vec{\mathbf{V}}\right)\right]_-^+\cdot\vec{\mathbf{N}}=0$$

– *Electric charge*:

$$\rho_a\frac{d_S z_S}{dt}+\nabla\cdot\mathbf{i}_a+\left[\mathbf{i}+\rho\left(z-z_S\right)\left(\mathbf{v}-\mathbf{V}\right)\right]_-^+\cdot\mathbf{N}=0$$

– *Entropy:* the questions of entropy balance and behavior laws were only addressed in the appendices to this chapter (section 5.6.3) in the context of the TPI. See also the example of application to a particular plane interface presented in Appendix 4 (section 5.6.4).

5.5. Conclusion

We first recalled the volume balance equations of fluid mixtures in the presence of an electromagnetic field[4]. We then considered the balance equations at the interfaces of continuous media in the presence of an electromagnetic field by expressing them in the Minkowski spacetime. In the case of a mixture, global electromagnetic quantities (not specific to each chemical species) were adopted. We limited ourselves to the case of low speeds v/c << 1. The presentation in four-dimensional space has advantages because it is very condensed, in particular for electromagnetic quantities, but it can also allow us to deal with relativistic cases.

5.6. Appendices

5.6.1. *Appendix 1: presentation of the Lorentz transformation*

A point M in spacetime is represented by a complex vector $_4\mathbf{M}$, whose associated row matrix $4\mathbf{M}^T$ is: $4\mathbf{M}^T = [x\ y\ z\ ict]$, where c is the speed of light in a vacuum in the absence of a field. The coordinates are indicated here in a Galilean frame of reference.

If we want to express these coordinates in another Galilean frame of reference of relative speed $\vec{\mathbf{v}}$, with $\mathbf{v}^T = [0\ 0\ v]$ with respect to the first frame of reference, we write: $4\mathbf{M'}^T = [x'\ y'\ z'\ ict']$, and the relation of change of reference is made by means of the special Lorentz transformation, linear transformation: $_4\mathbf{M}' = {_4\mathbf{D}} \cdot {_4\mathbf{M}}$, of matrix $\underline{\underline{4\mathbf{D}}}$:

$$\underline{\underline{4\mathbf{D}}} = \begin{bmatrix} 1 & 0 & 0 & 0 \\ 0 & 1 & 0 & 0 \\ 0 & 0 & 1/\alpha & iv/c\alpha \\ 0 & 0 & -iv/c\alpha & 1/\alpha \end{bmatrix}$$

with: $\alpha = \sqrt{1 - v^2/c^2}$.

4 It should be noted that the approach of de Groote and Mazur (1984) adopted here is partly called into question by Brechet et al. (2012) who believe they have identified a contradiction concerning the expression of the Lorentz force used.

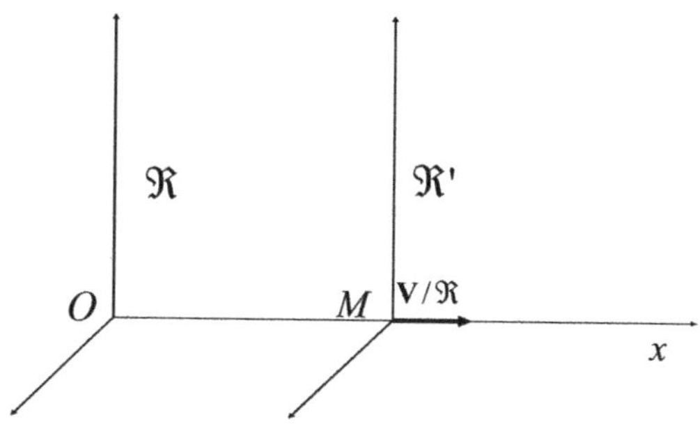

Figure 5.6. *Galilean reference frames. The reference frame $\mathfrak{R}$ is the one in which we observe events. At point M in instantaneous speed* $\mathbf{v}$*, we associate at time t the reference frame* $\mathfrak{R}'$ *moving at constant speed* $\mathrm{V} = \mathbf{v}$

In three-dimensional space, speed $\mathbf{v}$ of a material point can be defined at any instant. To define speed as an intrinsic quantity (not depending on the frame chosen to measure it) in spacetime, it is necessary to define a proper time (Figure 5.6). At the instant t, the material point M moves at speed $\mathbf{v}(t)$ in the reference frame $\mathfrak{R}$.

Now consider the Galilean frame moving at constant speed $\mathbf{V}$ such that, at moment t, we have $\mathbf{V} = \mathbf{v}$. The time τ measured with respect to such a frame (there is one at any time) linked to the material point is called the proper time of this material point. Let us assume that the coordinates remain zero; we will have, in the new frame:

$$\begin{bmatrix} 0 \\ 0 \\ 0 \\ ic\tau \end{bmatrix} = \underline{\underline{\mathbf{4D}}} \begin{bmatrix} 0 \\ 0 \\ z \\ ict \end{bmatrix}, \text{ let: } \begin{cases} 0 = z - vt \\ c\tau = \dfrac{1}{\alpha}\left(-\dfrac{v}{c}z + ct\right) \end{cases}$$

We deduce the expression for the proper time of the material point: $\tau = \dfrac{t}{\alpha}\left(1 - \dfrac{v^2}{c^2}\right) = \alpha t$

Note that here v and α, M although variable, must be considered constant when dealing with the Galilean frame coinciding with M and having the same speed as at time t. The preceding relation is therefore in fact a differential relation linking the proper time to the time in the frame $\Re$ and must be written with all rigors: $d\tau = \alpha\, dt$.

The four-vector $4\mathbf{V}$ is indeed a four-vector of Minkowski space as can be easily verified (Barral and Miedzik 2011).

5.6.2. *Appendix 2: application of the Lorentz transformation into electromagnetic quantities*

Let us assign indices 0, v and V to the values taken in the frames of reference at rest, following the particle motion and following any motion (which could be that of the interfacial zone), respectively.

The Lorentz transformation applies to four-tensors $4\vec{\vec{\mathcal{F}}}$, $4\vec{\vec{\mathcal{F}}}^*$ and $4\vec{\vec{\mathcal{H}}}$ with the following formulas for fields and displacements where V can be replaced by v:

$$\vec{\mathbf{D}}_{\mathrm{V}} = \vec{\mathbf{D}}_0 + (1/c)\vec{\mathbf{V}}\times\vec{\mathbf{H}}_0,\quad \vec{\mathbf{E}}_{\mathrm{V}} = \vec{\mathbf{E}}_0 + (1/c)\vec{\mathbf{V}}\times\vec{\mathbf{B}}_0,$$
$$\vec{\mathbf{H}}_{\mathrm{V}} = \vec{\mathbf{H}}_0 + (1/c)\vec{\mathbf{V}}\times\vec{\mathbf{D}}_0,\quad \vec{\mathbf{B}}_{\mathrm{V}} = \vec{\mathbf{B}}_0 + (1/c)\vec{\mathbf{V}}\times\vec{\mathbf{E}}_0$$

For the components of the four-momentum-energy tensor, the fundamental relationship between the fluxes becomes:

$$\vec{\vec{\mathbf{P}}}_0 = \vec{\vec{\mathbf{P}}}_v + (\vec{\mathbf{v}}/c)\otimes\vec{\mathbf{E}}_v\times\vec{\mathbf{H}}_v = \vec{\vec{\mathbf{P}}}_{\mathrm{V}} + (\vec{\mathbf{V}}/c)\otimes\vec{\mathbf{E}}_{\mathrm{V}}\times\vec{\mathbf{H}}_{\mathrm{V}}$$
$$\vec{\mathbf{J}}_{\mathbf{E0}} = \vec{\mathbf{J}}_{\mathbf{Ev}} + \vec{\mathbf{v}}\,\mathbf{E}_{ch} = \vec{\mathbf{J}}_{\mathbf{EV}} + \vec{\mathbf{V}}\,\mathbf{E}_{ch}$$

5.6.3. *Appendix 3: the laws of behavior at the interface*

The problem *of behavior laws, in the volume and at the interface*, is quite complex, particularly in the presence of surfaces. On the one hand, it is necessary to know the state laws and, on the other hand, to establish the complementary laws.

In rarefied plasmas in the presence of electromagnetic fields, the equations of continuous media alone are no longer sufficient and characteristic lengths of microscopic origin such as the Debye length or the Larmor radius intervene. On the contrary, the fields act differently on each species, so there is room for microscopic theories (Barral and Miedzik 2011).

We can nevertheless try to use our macroscopic theory and the principles of irreversible thermodynamics (TPI) to obtain the complementary laws (De Groot 1969; De Groot and Mazur 1984; Kluitenberg 1973; Brechet et al. 2012).

To do this, we must first determine the rate of entropy production, and then reveal generalized flows and forces that will then be assumed to verify linear relationships with phenomenological coefficients. For capillarity, we use the theory of the second gradient (Barral and Miedzik 2011).

Otherwise, we use the analysis on a molecular scale. In the presence of electromagnetic fields, intervene characteristic lengths of microscopic origin such as Delay and Larmor lengths.

The local *equation of the entropy* has the following form:

$$\rho_a \frac{ds_S}{dt} + \vec{\nabla}_S \cdot \vec{\mathbf{J}}_{Sa} + \left[\frac{\vec{\mathbf{q}}}{T} - \sum_k \vec{\jmath}_{Dk} \frac{g_k}{T} + \rho\left(s - s_S\right)\left(\vec{\mathbf{v}} - \vec{\mathbf{V}}\right) \right]_-^+ \cdot \vec{\mathbf{N}} = \dot{W}_{Sa} \qquad [5.10]$$

The expression of the **rate** of *entropy production* $\dot{W}_{Sa}$ is deduced from the other balance sheet equations and the state laws. It is written as:

$$\begin{aligned}\dot{W}_{Sa} = {} & \vec{\mathbf{q}}_a \cdot \vec{\nabla}_S \left(\frac{1}{T_S} \right) - \frac{1}{T_S} \sum_k \vec{\jmath}_{Dka} \cdot \left(T_S \vec{\nabla}_S \left(\frac{g_{kS}}{T_S} \right) - \vec{f}_{ka} \right) - \frac{1}{T_S} \vec{\vec{\mathbf{\Pi}}}_a : \vec{\nabla}_S \otimes \vec{\mathbf{v}}_S \\ & - \sum_k \frac{g_{kS}}{T_S} \dot{W}_{ka} + \left[\left(\frac{1}{T} - \frac{1}{T_S} \right) \vec{\mathbf{q}} - \sum_k \left(\frac{g_k}{T} - \frac{g_{kS}}{T_S} \right) \vec{\jmath}_{Dka} - \frac{1}{T} \left(\vec{\mathbf{v}} - \vec{\mathbf{v}}_S \right) \cdot \vec{\vec{\mathbf{\Pi}}} \right. \\ & \left. + \frac{\rho}{T} \left(\left(T_S - T \right) s + \sum_k \left(g_{kS} - g_k \right) Y_k - \frac{1}{2} \left(\vec{\mathbf{v}} - \vec{\mathbf{v}}_S \right)^2 \right) \left(\vec{\mathbf{v}} - \vec{\mathbf{V}} \right) \right]_-^+ \cdot \vec{\mathbf{N}} + \dot{W}_{Sa}^{(ch)}\end{aligned} \qquad [5.11]$$

with the *part due to the fields* linked to the rate of internal energy production of these fields:

$$\dot{W}_{Ea}{}^{ch} = -\dot{W}_{\psi a} - \frac{\partial E a^{ch}}{\partial t} + \vec{\nabla}_S \cdot \vec{\mathbf{J}}_{Ea}{}^{ch} - \left[\vec{\mathbf{J}}_E{}^{ch} \right]_-^+ \cdot \vec{\mathbf{N}} + \vec{\mathbf{v}}_S \cdot \left(\vec{\nabla} \cdot \vec{\vec{\mathbf{P}}}_a{}^{ch} + \left[\vec{\vec{\mathbf{P}}}^{ch} \right]_-^+ \cdot \vec{\mathbf{N}} \right) \qquad [5.12]$$

containing a potential production term $\dot{W}_{\psi a}$.

5.6.4. *Appendix 4: application examples*

The work of shaping the entropy production rate remains to be done in the general case. It has been done in particular cases such as the Peltier effect (Dudeck 1982).

It has also been carried out, in the case of the metal–plasma interface, in the presence of surface reactions, using reactive schemes and chemical kinetic laws that are not necessarily linear (Dudeck 1987).

The geometries of the interfaces studied are most often planar.

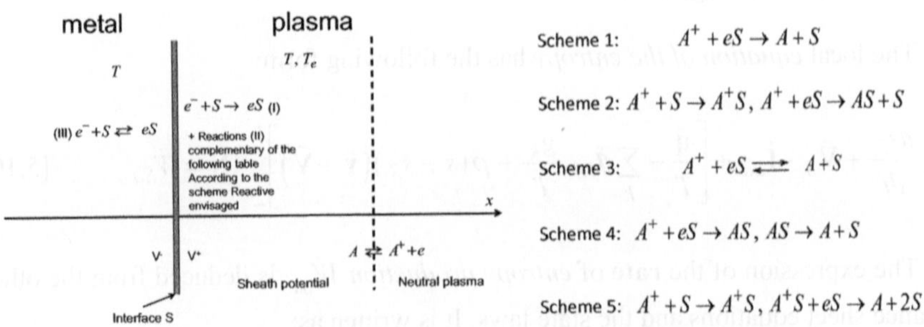

Figure 5.7. *Modeling of the metal-plasma interface under low pressure considered in the study of a Langmuir probe*

5.7. References

Barral, S. and Miedzik, J. (2011). Numerical investigation of closed-loop control for Hall accelerators. *Journal of Applied Physics*, 109(1), 013302.

Brechet, S.D., Reuse, F.A., Ansermet, J.-P. (2012). Thermodynamics of continuous media with electromagnetic fields. *The European Physical Journal B*, 85(412). doi: 10.1140/epjb/e2012-30719-4..

De Groot, S.R. (1969). *The Maxwell Equations: Non-Relativistic and Relativistic Derivations from Electron Theory*. North-Holland Publishing Company, Amsterdam.

De Groot, S.R. and Mazur, P. (1984). *Non Equilibrium Thermodynamics*. Dover Publications, New York.

Dudeck, M. (1982). Études thermodynamiques de phénomènes électromagnétiques interfaciaux. PhD Thesis, Université Pierre et Marie Curie, Paris.

Dudeck, M. (1987). Evolution thermique d'une sonde électrostatique dans un gaz raréfié et partiellement ionisé. *International Journal of Heat and Mass Transfer*, 30(3), 551–557.

Gatignol, R. and Prud'homme, R. (2001). *Mechanical and Thermodynamical Modeling of Fluid Interfaces*. World Scientific, Singapore.

Kluitenberg, C.A. (1973). On dielectric and magnetic relaxation phenomena and non-equilibrium thermodynamics. *Physica*, 68(1), 75–92.

Prud'homme, R. (2014a). *Flows and Chemical Reactions in Heterogeneous Mixtures*. ISTE Ltd, London and John Wiley & Sons, New York.

Prud'homme, R. (2014b). *Flows and Chemical Reactions in an Electromagnetic Field.* ISTE Ltd, London and John Wiley & Sons, New York.

6

The Interest of Microgravity for the Study of Fluid Interfaces

In this chapter, we address the potential interest that microgravity experiments can hold for fluid interface research. We must first identify the circumstances in which gravity is most likely to affect the interfaces. We will restrict ourselves to two scenarios: thin flames, which only exist outside of equilibrium but can be thought of as generalized interfaces, and capillary interfaces, where surface tension is present even at equilibrium. The available means for the experiment are briefly presented. Theory, experimentation and numerical computation are used to treat concrete examples.

List of symbols

$_{c,\, m}$	Resp. cut, maximum
Bo	Bond number
$f(\xi,t)$	Physical quantity $f = f_0 + f'$
f_0	Reference value
f'	Small disturbance of f
g	Acceleration of gravity
k	Wave number

Chapter written by Roger PRUD'HOMME and Kwassi ANANI.

L	Reference length
M	Mach number
p	Pressure
t	Time
U, V	Velocity
We	Weber number
x	Adimensional wave number
y	Adimensional amplification coefficient
γ	Isentropic coefficient
E	Small number in front of the unit
η	Interface rating
Ξ	Space variable
ρ	Density
σ	Superficial tension
φ	Speed potential
ω_i	Amplification coefficient

6.1. Introduction

The means of carrying out microgravity experiences have multiplied and been perfected over the past 50 years. The first free fall experiences carried out in the laboratory by Kumagai et al. (1970) with derisory means[1] made it possible to discover the law of linear regression of the area of drops in combustion and to confirm the theoretical laws. The construction of towers and weights of weightlessness made it

1 The free fall of the drop of fuel in combustion, 1 mm in initial diameter under a pressure of 1bar, could be observed for half a second.

possible to extend the duration a few seconds for a simulated weightlessness of 10^{-5}g and a better quality of the observations. Parabolic flights have become a common means within the reach of research teams in even longer times (approximately 20 seconds), but with a lower quality of microgravity (10^{-2}g). Quality and duration faults have been corrected with probes (10^{-5}g and several minutes), then with satellites – as currently the ISS – which allow a theoretically unlimited duration with an ideal quality of microgravity and accessibility experiences.

The experimenters on fluids in μg therefore have more and more possibilities, and the field of interfaces has largely benefited and continues to do so within the limits of the credits granted to the teams[2]. The areas concerning hair interfaces such as mosses and emulsions, as well as combustion interfaces and the behavior of near critical fluids are particularly interesting.

The situation of microgravity makes it possible to get rid of the parasitizes imposed by natural convection and thus to validate the laws of behavior of fluid systems often easier to theorize under these conditions. It allows us to provide possible events during space flights such as fires on satellite.

In this chapter, we will examine interface systems whose behaviors can be influenced by acceleration fields such as earthly gravity, and also that of other celestial bodies or by residual gravity in space station. The theoretical and digital results used to validate those that are connected to them will then be mentioned.

6.2. Instabilities between two bunk fluids in the presence of hair tension

6.2.1. *Presentation of the problem*

Study the effects of a small harmonic disturbance on two superimposed nonviscous fluids, with their separation surface initially horizontal, of respective density and speed (ρ_1, U_1), for the lower fluid and (ρ_2, U_2) for the upper fluid, and surface tension σ constant. The amplification factor ω_i disturbance is obtained from the analysis in small, linearized disturbances. We assume nonviscous fluids and semi-endless layers. Speed potential φ obeys the Laplace equation. Disruption φ' can be broken down[3] in normal

2 In France, the research teams concerned have mainly gathered for more than 30 years in a GDR (Research Group) CNRS-CNES, today known as GDR MFA (for "fundamental and applied microgravity").

3 In this linearized theory, each quantity $f(\xi,t)$ – with ξ space variable and time – is the sum of a reference value f_0 and disturbance $f = f_0 + f'$, $f' = \varepsilon\, f_0\, e^{i(k\xi - \omega t)}$ $\varepsilon \ll 1$, with real k and $\omega = \omega_r + i\omega_i$.

modes (Lamb 1945; Prud'homme 2013). The integration constants are determined using interfacial conditions. The interface equation is obtained using the function η, giving the rating of the interface disturbed compared to the reference plane surface. The limits at the interface (Laplace law for the pressure jump caused by the superficial tension and the conditions of waterproofing) are linearized, and the equations of the movement fluids can be expressed by the relations of Bernoulli. We obtain a homogeneous system of disturbance equations time with spatial derivatives of φ' and the function η. Finally, after the elimination of functions, we obtain the dispersion equation:

$$\left(\omega - k\frac{\rho_1 U_1 + \rho_2 U_2}{\rho_1 + \rho_2}\right)^2 = \frac{\sigma}{\rho_1 + \rho_2}k^3 + \frac{\rho_1 \rho_2}{(\rho_1 + \rho_2)^2}(U_1 - U_2)^2 k^2 + g\frac{\rho_2 - \rho_1}{\rho_1 + \rho_2}k \quad [6.1]$$

The amplification factor ω_i, imaginary part of the pulsation ω, deduces from the relationship [6.1]:

$$\omega_i^2 = -\frac{\sigma}{\rho_1 + \rho_2}k^3 + \frac{\rho_1\rho_2}{(\rho_1 + \rho_2)^2}(U_1 - U_2)^2 k^2 + g\frac{\rho_2 - \rho_1}{\rho_1 + \rho_2}k \quad [6.2]$$

where k is the number of waves and g is the acceleration of gravity[4].

Adimensionization

The previous equation can be put in adimensive form by introducing a reference length L and posing $x = kL$, dimensional wave number.

We then pose $y^2 = L^3 \frac{\rho_1 + \rho_2}{\sigma}\omega_i^2$, and we obtain the basic equation:

$$y^2 = Bo\,x + We\,x^2 - x^3 \quad [6.3]$$

with $Bo = \frac{gL^2(\rho_2 - \rho_1)}{\sigma}$, Bond number, and $We = \frac{\rho_1 \rho_2}{\rho_1 + \rho_2}\frac{(U_1 - U_2)^2 L}{\sigma}$, Weber number.

4 The real part obtained for ω is: $\omega_r = k(\rho_1 U_1 + \rho_2 U_2)/(\rho_1 + \rho_2)$.

6.2.2. *Rayleigh–Taylor instability*

Let us first consider the case of the Rayleigh–Taylor instability, where the fluids have equal velocities (*We* = *0*). The basic equation expressing the square of the reduced amplitude becomes:

$$y^2 = Box - x^3 \quad [6.4]$$

The instability of Rayleigh–Taylor can occur in the event that the surface tension is zero and the speeds of the two fluids are equal ($\sigma = 0, U_1 = U_2$), the upper fluid being the densest ($\rho_2 > \rho_1$). The flat surface is then unstable for any number of waves (for any speed of propagation).

On the contrary, if $\sigma \neq 0$ always with $U_1 = U_2$ and $\rho_2 > \rho_1$, the stability analysis gives a cut-off wavenumber k_c such that: $k_c^2 = (\rho_2 - \rho_1) g/\sigma$, which corresponds $x_c^2 = Bo$.

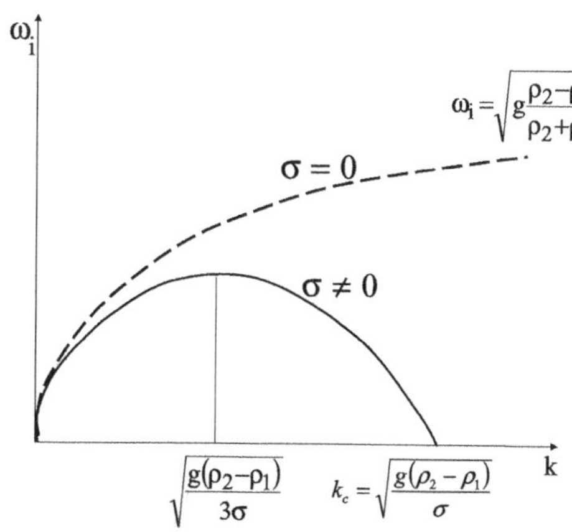

Figure 6.1. *Rayleigh–Taylor instability growth curves in nonviscous fluids when $U_2 = U_1$*

The Rayleigh–Taylor instability growth curves in nonviscous fluids when ($k > k_c$). The maximum appears for a particular value k_m wave number (see Figure 6.1) is given by: $k_m^2 = (\rho_2 - \rho_1) g/3\sigma$, i.e.:

$$x_m^2 = Bo/3, \text{and } y_m^2 = 2\left(\frac{Bo}{3}\right)^{3/2}.$$

Figure 6.2 shows Rayleigh–Taylor's instability. We have a glass of cold water on the surface of which we carefully inject hot colored water. This configuration is initially stable, but as the colored water cools and becomes denser than the water below, it becomes unstable (after Jens Niemeyer 2015).

Figure 6.2. *Rayleigh–Taylor instability produced by carefully injecting hot colored water in a glass of cold water. The falling "mushrooms" are characteristic. For a color version of this figure, see www.iste.co.uk/prudhomme/fluid3.zip*

6.2.3. *Kelvin–Helmholtz instability*

The instability of Kelvin–Helmholtz takes place with different speeds for both fluids. In the absence of gravity and superficial tension, the equation [6.2] becomes: $\omega_i^2 = \frac{\rho_1\rho_2}{(\rho_1+\rho_2)^2}(U_1-U_2)^2 k^2$. There is always a positive amplification coefficient, which indicates the instability in all values for the wave. The instability of Kelvin–Helmholtz classic gives rise to rolled structures, falling under nonlinear analysis, such as those represented in Figure 6.3(a).

These windings are often present in the compressible fluid mixture layers. It is possible to determine the convection speed U_{conv} large structures. The flow lines of the flow observed in a benchmark moving at speed U_{conv} are represented in Figure 6.3(b) and have structures[5] in throat (Papamoschou 1989).

5 These structures are similar to those observed for the dissemination flame against the tide and for the stretched plane premixed flame. This observation justifies the local use of these Lagrangian structures during the study and modeling of turbulent flames.

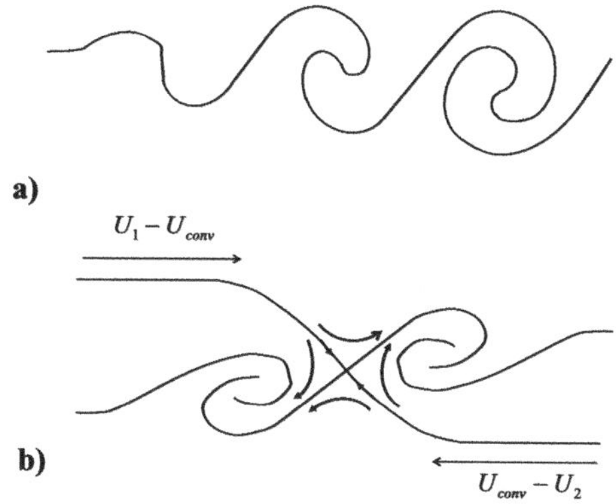

Figure 6.3. *a) Example of instability of Kelvin–Helmholtz which causes interface windings here in the event that $U_2 > U_1$. b) Current lines of a compressible fluid flow in the translation speed repository* $Uconv$

To determine the speed U_{conv} of a continuous flow of compressible fluid, we write that dynamic pressures p_0 of the two flows are equal to the stop point, i.e. that the reports $p_0/p = \left[1+(\gamma-1)M^2/2\right]^{\gamma/(\gamma-1)}$, where p is static pressure, γ is the isentropic coefficient and M is the number of Mach, are the same. If the numbers of Mach are not too high, and assuming isentropie, we get: $\rho_1(U_1 - U_{conv})^2 = \rho_2(U_{conv} - U_2)^2$.

If U_1 is the highest speed, we deduce[6]: $U_{conv} = U_1 \dfrac{1-(U_2/U_1)\sqrt{\rho_2/\rho_1}}{1+\sqrt{\rho_2/\rho_1}}$

Always with g null, but with a surface voltage of finished value, which is less classic, the amplification coefficient checks the relationship:

$\omega_i^2 = -\dfrac{\sigma}{\rho_1+\rho_2}k^3 + \dfrac{\rho_1\rho_2}{(\rho_1+\rho_2)^2}(U_1-U_2)^2 k^2$, which becomes in dimensional variables in this case where $Bo = 0$:

$$y^2 = -x^3 + Wex^2 = -x^2(x - We) \qquad [6.5]$$

6 This formula is no longer valid if the flow does not remain isentropic (shock waves attached to the structures may appear at high speeds). For a number of convection Mach $M_{conv} > 0.6$, the mixing layer observed experimentally becomes turbulent and has structures 3D.

The wave number of cutting $x_c = We$ appears in Figure 6.4, i.e.,

$$k_c = \rho_1\rho_2(U_1 - U_2)^2 / \sigma(\rho_1 + \rho_2)^2$$

and also a maximum rate of disruption growth for the reduced wave number (Prud'homme 2010) $x_m = 2We/3$, either: $k_m = 2\rho_1\rho_2(U_1 - U_2)^2 / 3\sigma(\rho_1 + \rho_2)^2$.

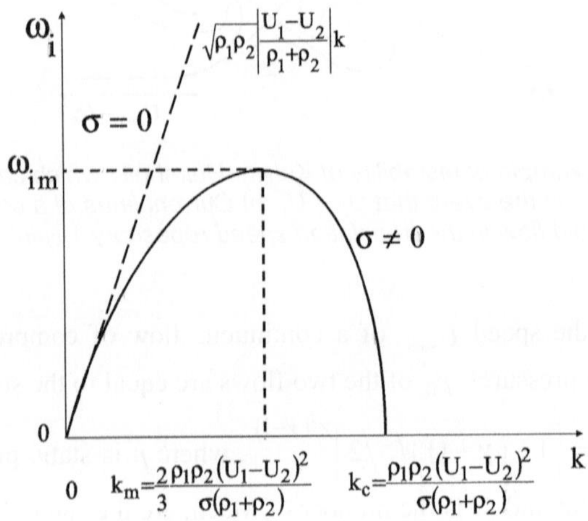

Figure 6.4. *Growth curves for an instability of Kelvin–Helmholtz when g = 0*

6.2.4. *The general case*

Let us analyze the general case characterized by the two horizontal layers in motion, i.e. with the numbers of Bond B_O and of Weber We of finished values.

Let us take up the dimensions without dimensions [6.3]: $y^2 = Bo\,x + We\,x^2 - x^3$ and try to determine characteristics of the evolution of the square of the amplitude of the disturbance y^2 according to the wave number x, as we did previously in sections 6.2.2 and 6.2.3.

The square y^2 of the amplitude as a function of x providing a third degree curve (Figure 6.5) includes an inflection point, with a part $x < 0$.

The two parties contain cut-off values and extrema. Only the positive x part is to be remembered.

The number of positive cut waves is equal to: $x_c^+ = \frac{We+\sqrt{We^2+4Bo}}{2}$, and the value to the maximum is: $x_m^+ = \frac{We+\sqrt{We^2+3Bo}}{3}$. We then find: $y^2{}_m{}^+ = \frac{Bo+(6Bo+2We)x_m^+}{9}$, and by replacing x_m^+ according to Bo and We, we find the expression $y^2{}_m{}^+(Bo,We)$.

In particular, we find the limits $y^2{}_m(0,We)=4We^2/27$, and $y^2{}_m(Bo,0)=\frac{Bo}{9}+\frac{2\,Bo^{3/2}}{3\sqrt{3}}$.

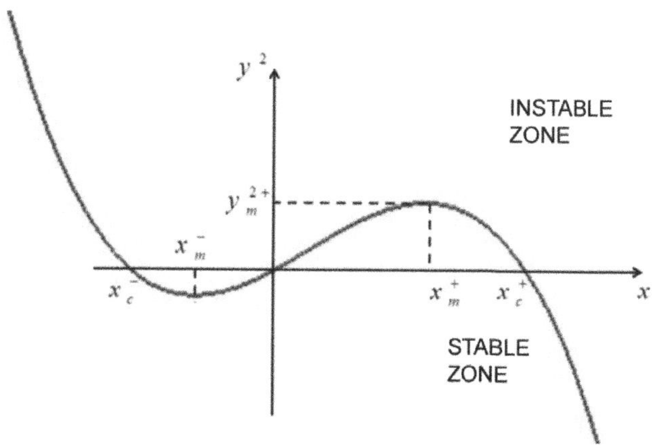

Figure 6.5. *General appearance of the growth curve of instability between two nonviscous fluids superimposed in dimensions without dimensions: square of the amplification factor y as a function of the x wave (only the positive x part is to be remembered). For a color version of this figure, see www.iste.co.uk/prudhomme/fluid3.zip*

Figure 6.6a shows the behavior of *a* and *b* functions on the instability of superimposed fluid layers.

Figure 6.6b represents the evolution of the product $x_c^+ y_m^{2+}$ supposed to characterize the intensity of instability for *Bo* between 0 and 2 and *We* between 0 and 3.

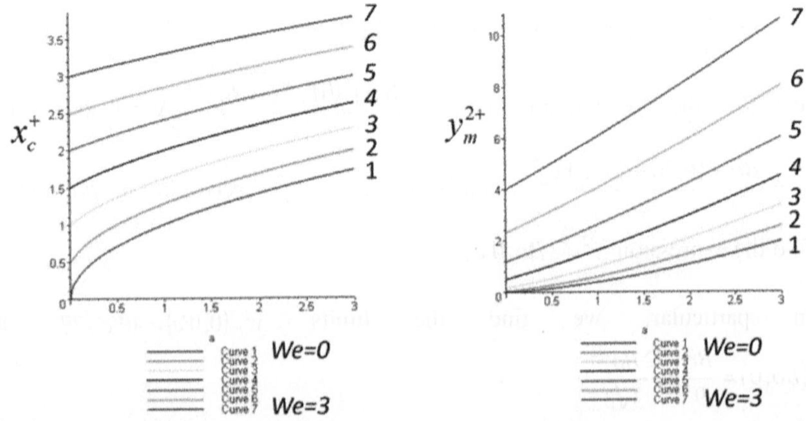

Figure 6.6a. *Influence of gravity (number of Bond Bo) on the instability of Kelvin–Helmholtz, i.e. the wave number of the unstable zone x_c^+ and the maximum amplification factor y^2m^+, for different values of the Weber We. For a color version of this figure, see www.iste.co.uk/prudhomme/fluid3.zip*

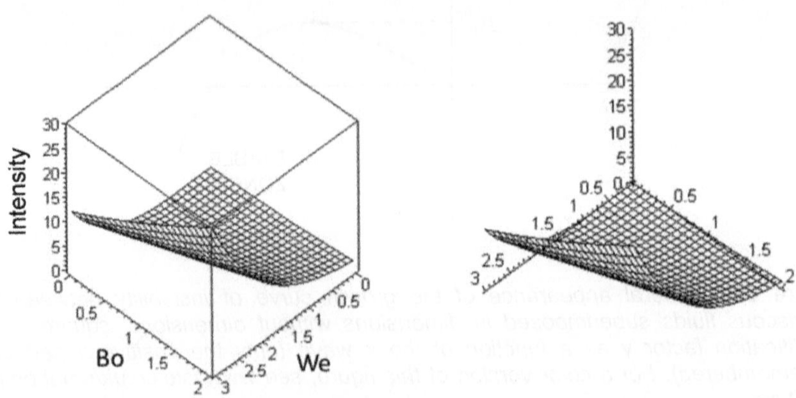

Figure 6.6b. *Influence of Bo and We on the "intensity" of the coupled instability resulting from Rayleigh–Taylor and Kelvin–Helmholtz*

6.3. Interest of the microgravity in the presence of capillarity: influence of gravity on the structure and dynamics of the foam

Figure 6.7 shows the nature of a diphasic aqueous medium with air bubbles as a function of the liquid fraction. We see the area occupied by the foams. The transition zone separates the foam with rather solid behavior from that with rather liquid behavior.

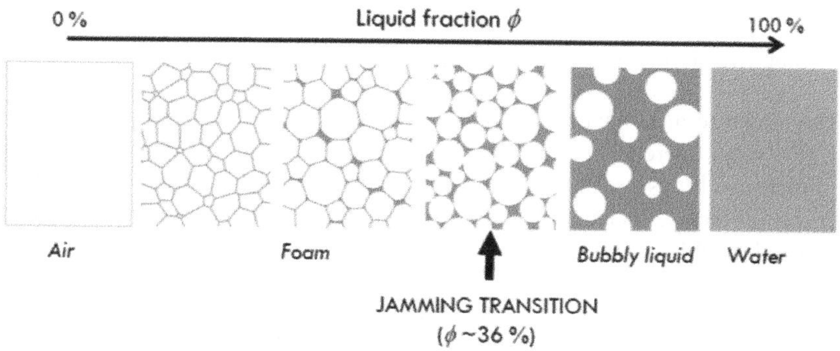

Figure 6.7. *Designation of an air/water diphasic structure as a function of the liquid fraction (Langevin 2017). For a color version of this figure, see www.iste.co.uk/prudhomme/fluid3.zip*

The foams (Monnereau et al. 1999; Weaire et al. 2001) are random dispersions of gas bubbles separated by thin liquid films. They constitute a model of cellular structure which is frequent in nature: the magnetic and crystalline systems of ceramics and alloys, the biological tissues, etc. They have had important applications from the extinction of fires to cell materials, passing through soft drinks. Another motivation for the study of foams is also the search for non-foamability or at least of its lessened. For example, in the tanks of capillary sponge satellites, the formation of stable bubbles in the fuel supply grid could be catastrophic.

The foam is an ephemeral object whose lifespan depends on the assessment of the forces present.

Destabilizing forces are:

– drainage of liquid films by gravity which leads to the coalescence of bubbles;

– disproportionate (ripening of Ostwald) during which the smallest bubbles are empty in the largest due to the hair pressure, then disappear.

Stabilizing forces are:

– long-range forces (electrostatic or steric);

– the Gibbs–Marangoni effect, which restores the initial thickness of the films when they have undergone local slimming, and which depends on the rheological properties of surface dilation of the foaming solution.

These effects are studied on a local scale while the user is generally interested in the macroscopic behavior of the foam. An essential problem for the study of foams concerns the transition from microscopic to macroscopic, an intermediate step concerning two-dimensional foams[7].

The structure and dynamics of these systems are relatively poorly understood. For a few years, there have been a certain number of studies for digital modeling of two-dimensional mosses for a few years, and only recently, has there been a start of modeling of three-dimensional foams. In general, experiences are not numerous enough and reliable to validate the models.

Any phenomenon studied on a liquid foam is always coupled with gravity drainage whose essential action is, on the one hand, to generate variations in the thickness of the films along the foam column and, on the other hand, to accumulate large bubbles (lightest parts) in its upper part. The purpose of our microgravity study is to analyze the evolution of the morphology of three-dimensional foam under the action of the forces of physico-chemical origin (capillary pressure, interfacial rheology, etc.) and also to study how a foam can interact with different solid walls simulating those of a satellite tank, for example (Prud'homme and Zappoli 1995; Langevin 2017).

The team has proceeded to many experiences in a probe rocket. Others have been envisaged for the ISS.

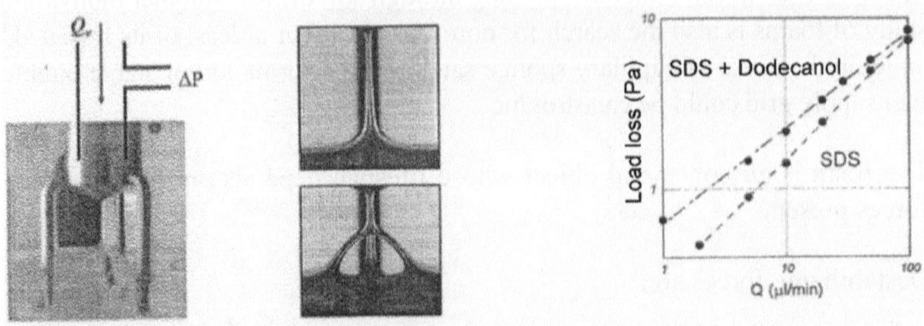

Figure 6.8. *Measurement of the load loss of a flow in the board of the plateau and a node for a solution of SDS, a mixed solution of SDS and dodecanol*

Example of experience provided to study the transport of liquid and particles in the edges of the tray:

> Modelization exists based on the analysis of viscous dissipation in the foam. In fact, physics is much more complex. To isolate the role of surface rheology, experiences are planned on the ISS on dry and

7 The two-dimensional foam is created by blowing nitrogen into a soapy solution in a gap between two plates spaced 1–2mm apart.

humid foams (Group of Langevin (Safouane et al. 2001) in Orsay). In this context, they have been developed in the laboratory (LPMDI) of experiences on the transport of liquid in the plateau network (Figure 6.8) (Prud'homme and Zappoli 2003).

Figure 6.9 (Prud'homme and Zappoli 1995) shows an example of foam drying. The foam created at 0G (Figure 6.9a) was made up of perfectly spherical bubbles and turned under 1.8g (Figure 6.9b) into a set of polygonal structures.

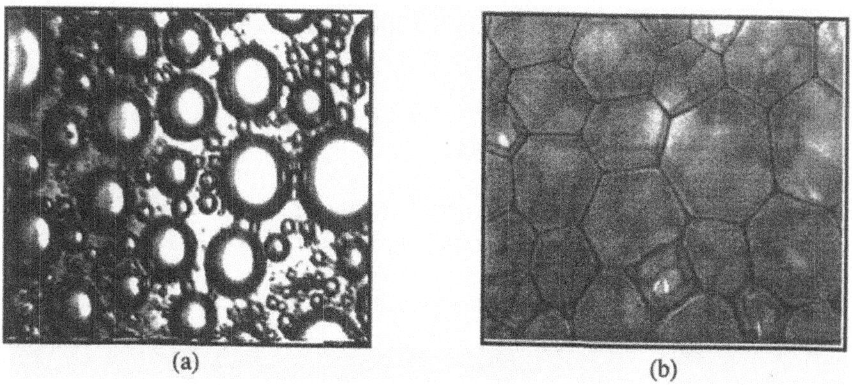

Figure 6.9. *Image of a foam (a) formed in microgravity and (b) dry under 1.8g (Prud'homme and Zappoli 1995)*

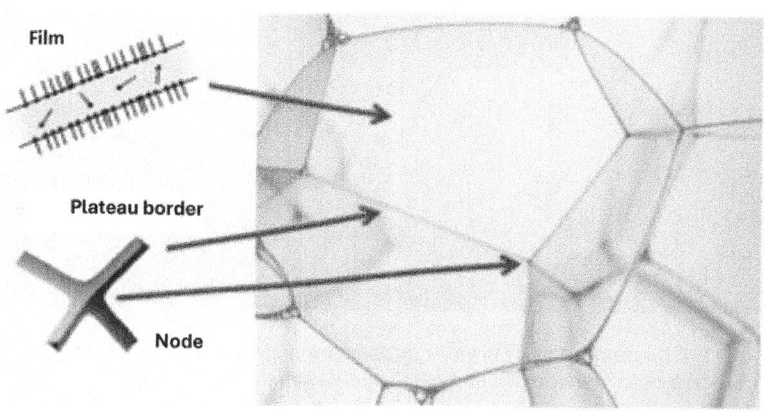

Figure 6.10. *Aqueous dry foam stabilized by tension active molecules represented by a circle (polar head), in contact with water, and a hydrophobic chain, in contact with air (Langevin 2017). For a color version of this figure, see www.iste.co.uk/prudhomme/fluid3.zip*

Langevin studied the stabilizing action of surfactants on mosses related to microgravity experimentation (Langevin 2017). It highlighted the importance of surface rheology and confinement in thin foam films for the behavior of surfactants, even in weightlessness. Figure 6.10 shows a diagram of aqueous dry foam with liquid films separating the sides of the bubbles, the borders of the tray and the nodes visible to their corners.

In addition to the foams, many other physical systems involve hair interfaces. Some – such as mooring phenomena (Caltagirone 2023), bubble flows (Colin et al. 1996) and emulsions (Antoni and Steinchen 2001) – do, or have been, the subject of microgravity studies.

6.4. Influence of gravity on flames

Various combustion problems reveal more or less sensitive flames to acceleration fields (Durox et al. 1989) and, in particular, gravity.

Figure 6.11 gives a blatant example of this influence on a candle flame[8].

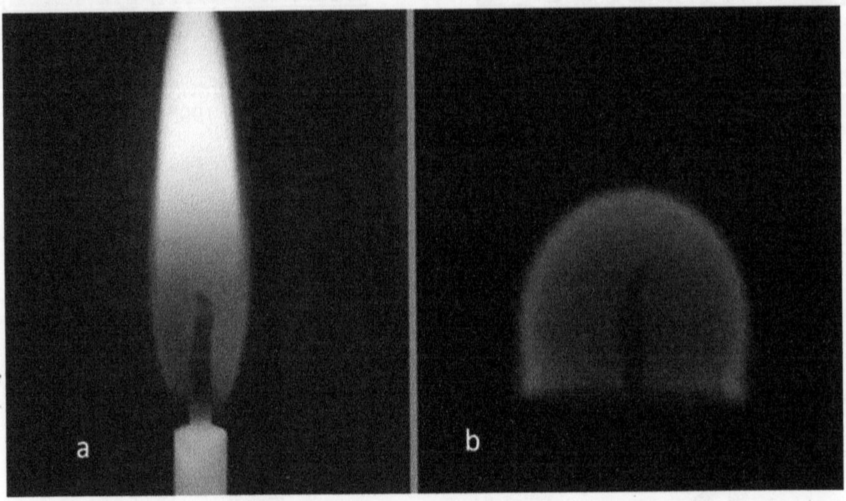

Figure 6.11. *Comparison of a flame a) under earthly gravity and b) weightlessness in the ISS. For a color version of this figure, see www.iste.co.uk/prudhomme/fluid3.zip*

Figure 6.12 shows an air–methane premixture flame above a Bunsen spout, observed in two situations in parabolic flight. As a micro-peak, the flame is calm,

8 See: https://commons.wikimedia.org/wiki/File:Candlespace.jpg.

floating a little. During a resource causing an acceleration of twice the normal severity, the summit undergoes periodic vacillating[9] (Qian et al. 2024) faster than under normal severity. These fluctuations are explained by the instability of the type of Kelvin–Helmholtz[10] due to the jet of burnt gas rushing faster than the surrounding fresh gases.

When we study the combustion of drops in a spray flow, it is generally assumed that the gas movement in the vicinity of each droplet is with spherical symmetry (Anani et al. 2023). This is reality if the droplets are small enough to be trained by the comburant fluid. But this is default for large drops due to the appearance of relative speed. It is important to know the phenomena likely to destabilize spherical drops in combustion in a disturbed atmosphere (Mauriot and Prud'homme 2014).

Figure 6.13 reflects the effect of the relative convection on the distribution of the mass fractions of the constituents during the combustion of sodium drops in the moving atmosphere (Rao et al. 2012). These ruptures of spherical symmetry caused by forced convection can also appear during the combustion of a drop to rest due to the natural convection induced by the heat released (Prud'homme 2022).

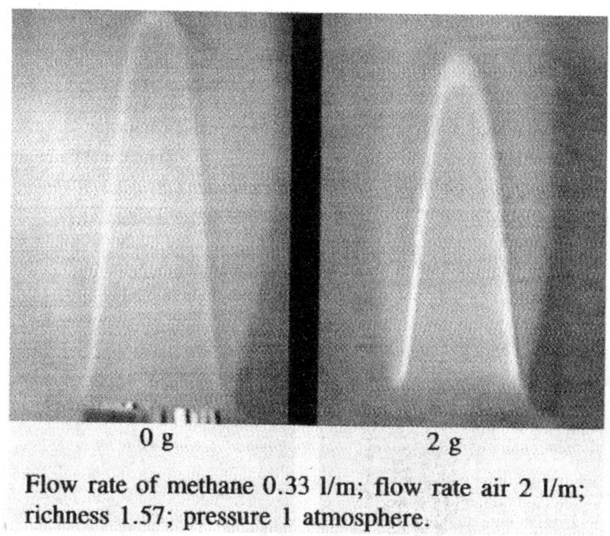

Figure 6.12. *Flame prevented two moments – 0g and of 2g resource – of a parabolic flight. For a color version of this figure, see www.iste.co.uk/prudhomme/fluid3.zip*

9 Periodic flickering corresponds to the whitish zone visible on 2g photography at the top of the flame.

10 With the difference that gas jets have accelerated movements above the burner, while the fluid layers are at constant speeds in the study of Kelvin–Helmholtz.

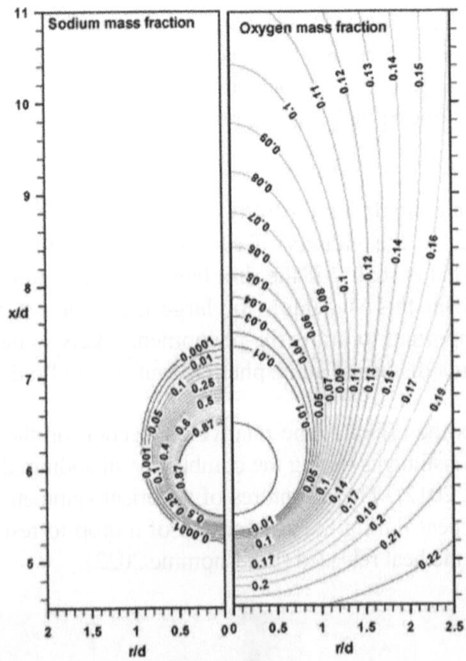

Figure 6.13. *Iso-fraction curbs sodium and oxygen around a drop. For a color version of this figure, see www.iste.co.uk/prudhomme/fluid3.zip*

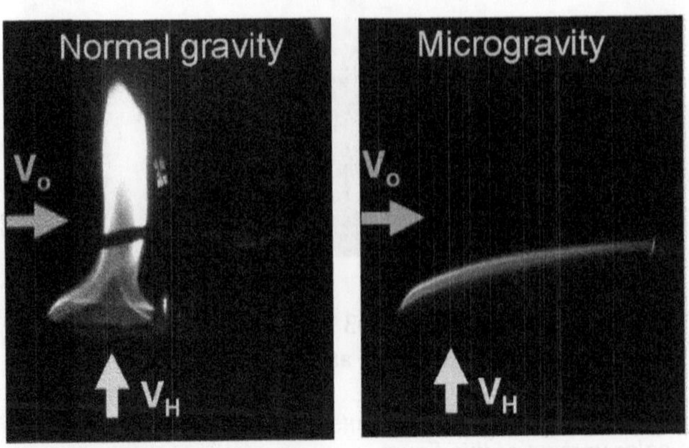

Figure 6.14. *Experimental observations of a broadcasting flame under normal gravity and under microgravity (Prud'homme 2010). For a color version of this figure, see www.iste.co.uk/prudhomme/fluid3.zip*

Fighter can induce such natural convection movements when combustion of plates or stems. In the experiences of simulation in Figure 6.14, the gas fuel is injected at V_H speed through a porous plate and the comburant circulates in parallel with the V_O speed (Prud'homme 2010).

Among the combustion problems requiring microgravity studies, we can also cite interactions between drops (Chauveau et al. 2006) or stems (Citerne et al. 2014.) in simultaneous combustion (Figure 6.15) (Prud'homme et al. 2017).

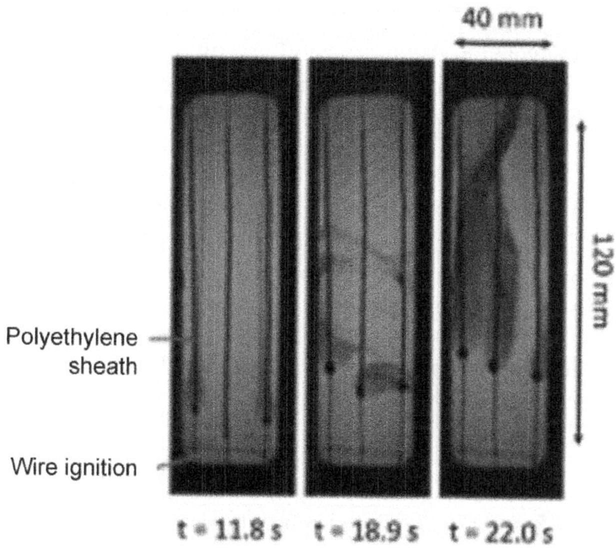

Figure 6.15. *Images obtained using the absorption technique at three instants after the start of the microgravity period (t = 0) (Prud'homme et al. 2017). The air flow and flame propagation are oriented from bottom to top*

6.5. Conclusion

We have only touched them, but the situations mentioned in this chapter illustrate the possibilities offered by the microgravity experiment obtained by mechanical compensation to study the fluid systems with interfaces.

Other means are sometimes used in the ground experimentation to eliminate gravity. Hele–Shaw configurations[11] are limited to systems with two-dimensional behavior. Sound wave lift is sometimes used. Another method is magnetic lust.

11 A Hele–Shaw cell consists of two very close glass plates of each other between which one or more fluids are injected. See: https://fr.wikipedia.org/wiki/Cellule_de_Hele-Shaw.

These means sometimes have an interest due to greater accessibility, and also due to lower costs than spatial experimentation.

We have shown in section 6.2, how the acceleration of gravity could influence the instabilities of a physical system simply made up of two superimposed mobile horizontal layers separated by a hair interface. Only the nonviscous case was presented. Let us quote a problem in a cylindrical configuration recently studied by Gouin (2020) with a concrete application to the behavior of plants under the effect of lunisolar tides.

Section 6.3 is devoted to the influence of gravity on the structure and dynamics of mosses, limiting itself to rather descriptive general characteristics. We have been cited by other phenomena involving hair interfaces such as mooring, bubble flows and emulsions.

Systems with generalized interfaces existing in the field of combustion have been presented in section 6.4: flames of flickering premises under 1g and 2g becoming calm in µg, combustion of drops and combustion of plate interactions between stems in simultaneous combustion.

The applications of this research have been highlighted. Relatively supplied documentation accompanies these presentations, giving the reader more information for their reading.

Finally, it should be noted that the study of other systems including a fluid phase, but in contact with a solid phase, has also benefited for years from the possibilities offered by the microgravity experiment. This is the case of research carried out on crystallization and the solidification of metals and metal alloys by the teams of the GDR MFA, as evidenced by a work quoted in (Prud'homme et al. 2001). Five other publications concerning the work of this GDR are also cited (Zappoli and Prud'homme 1999; Prud'homme and El Ganaoui 2004a; Prud'homme and El Ganaoui 2004b; Etay et al. 2017; Beysens et al. 2007) in the references.

6.6. References

Anani, K., Prud'homme, R., Hounkonnou, M.N. (2023). An approximate analytical model for evaporating droplets frequency response in a mixed injection regime. *Comptes rendus mécanique*, 351(S2), 77–95. doi: 10.5802/crmeca.146.

Antoni, M. and Steinchen, A. (2001). Numerical approach to phase separation in a model emulsion. *Phys. IV France*, 11(Pr6), 281–288.

Beysens, D., Chatain, D., Garrabos, Y., Lecoutre, C., Palencia, F., Evesque, P., Nikolayeva, V. (2007). The effect of vibrations on heterogeneous fluids: Some studies in weightlessness. *Acta Astronautica*, 61(11–12), 1002–1009.

Caltagirone, J.-P. and Prud'homme, R. (2023). Modélisation du mouillage partiel par une approche macroscopique en mécanique discrète. *ISTE Open Science*, 6(1). doi: 10.21494/ISTE.OP.2023.1019.

Chauveau, C., Halter, F., Gökalp, I. (2006). Vaporization in three-dimensional droplet arrays: Effects of the fuel vapor saturation. In *10th International Conference on Liquid Atomization and Spray Systems*, Kyoto.

Citerne, J.M., Dutilleul, H., Kizawa, K., Fujita, O., Kikuchi, M., Jomaas, G., Legros, G. (2014). Propagation de flamme sur des échantillons de révolution cylindrique : premiers essais en micropesanteur. In *Colloque annuel du GdR MFA*.

Colin, C., Fabre, J., Mcquillen, J.B. (1996). Bubble and slug flow at microgravity conditions: State and knowledge and open questions. *Chem. Eng. Comm.*, 141–142(1), 155–173.

Durox, D., Prud'homme, R., Scouflaire, P. (1989). Premixed flames in microgravity. Combustion experiments during KC-135 parabolic flights. Document, ESA SP 1113, Paris.

Etay, J., Prud'homme, R., Haldenwang, P. (eds) (2017). Basic and applied researches in microgravity – A tribute to Bernard Zappoli's contribution. *Comptes rendus mécanique*, 345(1), 1–116.

Gouin, H. (2020). Influence of lunisolar tides on plants. Parametric resonance induced by periodic variations of gravity. *Phys. Fluids*, 32(10), 101907. doi: 10.1063/5.0023717.

Kumagai, S., Sakai, T., Okajima, S. (1970). Combustion of free fuel droplets in a freely falling chamber. In *13th Symposium on Combustion*, Salt Lake City.

Lamb, H. (1945). *Hydrodynamics*. Cambridge University Press, Cambridge.

Langevin, D. (2017). Aqueous foams and foam films stabilized by surfactants. Gravity-free studies. *Comptes rendus mécanique*, 345(1), 47–55. doi: 10.1016/j.crme.2016.10.009.

Mauriot, Y. and Prud'homme, R. (2014). Assessment of evaporation equilibrium and stability concerning an acoustically excited drop in combustion products. *Comptes rendus mécanique*, 342(4), 240–253.

Monnereau, C., Vignes-Adler, M., Kronberg, B. (1999). Influence of gravity on foams. *J. Chim. Phys.* 96(6), 958–967.

Niemeyer, J. (2015). Rayleigh-Taylor instability. *YouTube* [Online]. Available at: https://youtu.be/yabqo7VFTYs.

Papamoschou, D. (1989). Structure of the compressible turbulent shear layer. In *27th Aerospace Science Meeting*, Reno.

Prud'homme, R. (2010). *Flows of Reactive Fluids*. Birkhäuser, Boston.

Prud'homme, R. (2013). *Flows and Chemical Reactions in Homogeneous Mixtures*. ISTE Ltd, London and John Wiley & Sons, New York.

Prud'homme, R. (2022). Instabilities in a spherical liquid drop. *Front. Space Technol.*, 3, 835464. doi: 10.3389/frspt.2022.835464.

Prud'homme, R. and El Ganaoui, M. (eds) (2004a). *Microgravité et transferts – Numéro thématique de mécanique et industries*. EDP Sciences, Les Ulis.

Prud'homme, R. and El Ganaoui, M. (eds) (2004b). Microgravity. *Comptes rendus mécanique*, 332(5–6), 319–486.

Prud'homme, R. and Zappoli, B. (1995). Rapport d'activité 1994–1995 du groupement de recherches "Mécanique des fluides et phénomènes de transport en microgravité" (GDR 1028). Technical Report, Laboratoire de Modélisation en Mécanique.

Prud'homme, R. and Zappoli, B. (2003). Bilan à 4 ans 2000–2003 du GDR 2258. Technical Report, Laboratoire de Physique des Matériaux Divisés et des Interfaces (LPMDI), Université de Marne-la-Vallée, Marne-la-Vallée.

Prud'homme, R., Langevin, D., Faivre, G. (eds) (2001). Sciences de la matière en microgravité. *Phys. IV France*, 11(Pr6), 109–118.

Prud'homme, R., Legros, G., Torero, J.L. (2017). Combustion in microgravity: The French contribution. *Comptes rendus mécanique*, 345(1), 86–98.

Qian, C., Yang, Y., Wang, G., Krikunova, A., Hu, K. (2024). Effect of gravity orientation on flickering characteristics of premixed conical flames. *Microgravity Science and Technology*, 36(3). doi: 10.1007/s12217-023-10088-3.

Rao, P.M., Raghavan, V., Velusamy, K., Sundararajan, T., Shet, U.S.P. (2012). Modeling of quasi-steady sodium droplet combustion in convective environment. *International Journal of Heat and Mass Transfer*, 55(4), 734–743.

Safouane, M., Durand, M., Saint Jalmes, A., Langevin, D., Bergeron, V. (2001). Aqueous foam drainage. Role of the rheology of the foaming fluid. *Phys. IV France*, 11(Pr6), 275–280.

Weaire, D., Cox, S.J., Hutzler, S., Bradley, G. (2001). Foams in microgravity. *Phys. IV France*, 11(Pr6), 213–220.

Zappoli, B. and Prud'homme, R. (eds) (1999). *Journées d'Oléron 1998 du GDR PR2M*. EDP Sciences, Les Ulis.

7

Fire–Air: A Story of the Flame

We relate how at the end of the 18th century, a discovery linked to combustion, challenged the idea of what we had regarding constituents of matter and led to revolutionizing chemistry. A correct interpretation of this discovery had to be found and for that to question existing theories and dogmas. It then took a fundamental reflection on experimental work and the importance of quantitative measurements, to invent a scientific way of operating. The innovations and the progress made during the following two centuries in physics and chemistry, as well as in thermodynamics and mechanics, has brought us to the current situation where combustion has found its place as a science in its own right.

7.1. Introduction

The fascination with fire has led humans to give it a fundamental place in the explanation of nature[1,2].

Chapter written by Roger PRUD'HOMME, Mahouton Norbert HOUNKONNOU and Guillaume LEGROS.

1 It is the mastery of fire which would have allowed the human species to dedicate calories to brain activity rather than to maintaining its body temperature. This, in some way, links the fire element to the development of knowledge.

2 Also, according to https://pubmed.ncbi.nlm.nih.gov/33230523/, one of the major genetic innovations of Sapiens would be a gene that made it less prone to cancers, developed due to the inhalation of fumes from domesticated fire; a gene that Neanderthals had not developed. This would link fire to the survival of our species! Another equally vital question is that of the responsibility of societies in the control and use of fire (climate change, major fires, warlike destruction, etc.).

For hundreds of years, fire was considered one of the four elements that make up matter, along with air, earth and water.

Fire only lost its position as a basic element when it was understood that it was only the result of globally exothermic chemical reactions of oxidation–reduction between molecular constituents. This was possible from the moment when the basic rules of combustion were understood, thanks to the development of modern chemistry initiated by Lavoisier (1743–1794)[3].

The basic elements – liquid, solid or gaseous – became atoms again, as ancient philosophers had long suspected, and molecules resulted from their combinations. Note that these atoms are themselves divisible, contrary to what the atomic philosophers thought[4].

Progress in phase change physics and its thermodynamic component complemented that of chemistry.

It took time and insight to establish a coherent classification of atomic elements and open a new field of chemistry, the laws of physics and mechanics complementing knowledge of combustion and many other areas of science. Our knowledge evolves in this way, with stagnations and revolutions.

The field of possibilities then vastly expands, and that of achievements depends on the awareness that societies and individuals have of them. We can bet that there will be many other revolutions in science and that human beings will make the right choices of applications while being aware of the risks, in order to avoid catastrophes and economic, ecological and social disasters.

In this chapter, we attempt to take stock of the discoveries and conceptual developments that have led to the current situation of combustion and its interfaces, which are flames. This work can usefully complement Appendix 2 of a relatively recent article (Prud'homme 2019).

---

3 The main foundations of the Chemical Revolution initiated by Lavoisier are the law of conservation of mass (Lavoisier's law: "Nothing is lost, nothing is created, everything is transformed") and the theory of combustion by oxygen.

4 The atomistic theory supported the idea of matter composed of indivisible "grains" (against the idea of matter indefinitely divisible). It was notably defended by Leucippus and his disciple Democritus (c. 460–370 BCE), philosophers of ancient Greece, as well as in India, further back in time. In fact, atoms are not indivisible but are themselves made up of subatomic particles: protons, neutrons and electrons, as well as their constituents (including quarks).

7.2. The phlogiston theory

Phlogiston theory is a chemical theory that explained combustion by postulating the existence of a “flame element”, a fluid called φλογιστόν (phlogiston) (from the Greek φλόξ phlox, flame), present within combustible bodies. It was conceived by Becher (see Figure 7.1(a)) in the late 17th century and developed by Georg Ernst Stahl (see Figure 7.1(b))[5].

a) b)

Figure 7.1. *a) Johann Joachim Becher (1635–1682) who was an alchemist, doctor and German chemist. b) Georg Ernst Stahl (1659–1734) who was a chemist and doctor*

It was the alchemist, chemist and physician Johann Joachim Becher who proposed the theory of phlogiston. The basic idea is that, since flames are obviously real, there must necessarily be an element that participates in the constitution of combustible bodies, fluid – as flames obviously are – and which, following combustion, would be the principal concrete constituent of these flames.

5 For his work in chemistry, Stahl was the most acclaimed and most hated scientist of the 18th century. In his time, he was idolized throughout Europe. The article “Chemistry” in Diderot’s Encyclopedia presents him as the equal of Newton, achieving the chemical equivalent of Newton in physics. However, soon, Lavoisier founded modern chemistry by refuting a concept of Stahl, phlogiston or phlogiston.

The idea is related to the four elements of *Empedocles*[6] which are supposed to constitute everything according to natural philosophy (see Figure 7.1(c)).

Figure 7.1(c). *The diagram of Encephloque. The four elements: air, fire, earth and water. For a color version of this figure, see www.iste.co.uk/prudhomme/fluid3.zip*

The four elements discussed come from the observation that to grow a seed, we need earth, water, air and fire (in the sense of sun, energy).

This composition has become obvious, true at all times.

Today, there is a deeper understanding of it. In particular, it has been discovered that fire is formed by a chemical reaction[7], and does not, in reality from a phenomenological point of view, oppose fire as an element in the sense of sun or energy.

6 Empedocles (-490 to -430 BCE) is a Greek philosopher, poet, engineer and physician from Sicily, from the 5th century BCE. He is one of the pre-Socratic philosophers, those first thinkers who attempted to discover the archè of the cosmos, its “schema”.

7 As we will see it below.

7.3. The discovery of "air to fire"

In August 1774, Joseph Priestley (see Figures 7.2(a) and (b)) isolated an "air" which appeared to be of an unknown type[8], but he did not have time to continue studying it, having to leave for a tour of Europe with Shelburne.

a)

b)

Figure 7.2. *a) Replica (on a reduced scale) of the burning glass used by Priestley for the discovery of oxygen, exhibited in the laboratory of the Joseph Priestley House in the United States[9,10]. b) Joseph Priestley (1733–1804) (portrait 1794). Theologian, dissenting priest, natural philosopher, pedagogue and political theorist. For a color version of this figure, see www.iste.co.uk/prudhomme/fluid3.zip*

Shortly afterward, Carl Wilhelm Scheele (see Figure 7.2(c)) produced the same kind of air in Sweden and proved its presence in the atmosphere. He also found that this gas disappeared when a body was burned in it or when a body was burned in ordinary air; he therefore called it "fire air". Scheele (born December 1742 in Stralsund in Swedish Pomerania and died May 21, 1786, in Köping) was a

8 Priestley produced a kind of air "which he found so pure or free from phlogiston that, by comparison, ordinary air already seemed vitiated". He called it dephlogisticated air.

9 The "Joseph Priestley House" is the American home of the 18th-century British theologian, dissenting priest, natural philosopher, educator and political theorist Joseph Priestley (1733–1804). He lived there from 1798 until his death in 1804.

10 A large convex lens that allows the sun's rays to be concentrated on a small surface, heating it and causing it to burn. That is, a magnifying glass!

Swedish–German chemist. He is particularly known for having discovered many chemical elements and compounds, such as oxygen and chlorine. An outstanding chemist, he had the particularity of using rudimentary instruments (in particular, he collected gases in pigs' bladders).

Figure 7.2(c). *Carl Wilhelm Scheele (1742–1786) was a chemist, pharmacist and pharmacologist*

Priestley and Scheele had both produced oxygen, but without knowing what they had at hand. They "were unable to free themselves from the phlogiston categories" as they found them established. The element that was to overturn the whole phlogiston conception and revolutionize chemistry remained, in their hands, stricken with sterility.

7.3.1. *Oxygen*

> But Priestley had immediately communicated his discovery to *Lavoisier* (Grimaux 1888) (see Figure 7.3), in Paris, and the latter, starting from this new fact, subjected the whole of phlogiston chemistry to investigation; he was the first to discover that the new kind of air was a new chemical element, that, in combustion, it is not the mysterious phlogiston that escapes, but rather this new element, which combines with the body; and he thus put on its feet the whole of chemistry which, in its phlogiston form, was turned upside down. And if it is not true, contrary to what he later claimed, that he produced oxygen at the same time as Priestley and Scheele and independently of them, he nevertheless remains the one who really discovered oxygen

> compared to the other two, who had simply produced it, without having the slightest idea of what they had produced[11].

From then on, the material field of transformations was reducible in law to the equations that describe the occurrences of the combinations, which entails the correlative obligation to close any experimental system, in order to maintain its exact quantitative definition.

Lavoisier would lead by example in this requirement of closure which would have immense consequences in the theory of natural phenomena: will thermodynamics not be constituted as a body of descriptive relations of isolated systems?

It is also by virtue of his principle of economy that Lavoisier noted the uselessness of giving phlogiston a mediating function and introduced the effective action of oxygen.

This would peg, but imperfectly, the elementary distribution of substances, according to a didactic as well as cosmological project. Its fundamental contribution, ultimately, would have been to link in an effort of representation, at once scrupulous and global, a method of rational nomenclature, a methodical classification and a theory of matter where everything is resolvable in rearrangements.

Lavoisier did not in any way enrich the inventory of natural things, nor did he discover any phenomenon that had been unknown before him. However, we owe him, on the other hand, for having conceived and tested a method of thinking about the representation of the material universe. Thanks to him, the relationship between substance and noun will henceforth be thought of by means of weighing. The analytic must find its exact image in the denotative, and this accounts for the associative structure of combinations through the vocalic composition. The consistency of his system is based on the Newtonian principle of economy:

> Nothing is created, neither in the operations of art, nor in those of nature, and we can posit as a principle that, in any operation, there is an equal quantity of matter

11 This text is taken from the Preface by Friedrich Engels to Book II of Marx's Capital (Editions sociales, translation by Erna Cogniot, 1960). Engels takes the example of the discovery of oxygen to explain its revolutionary aspect. He then applies these considerations to the theory of surplus value set out in Karl Marx's famous work "Capital" by stating: "Marx is to his predecessors (such as Ricardo and Rodbertus, in economics, editor's note) as regards the theory of surplus value, what Lavoisier is to Priestley and Scheele (as regards the chemical revolution, editor's note)".

before and after the operation; that the quality and quantity of the principles are the same and that there are only changes, modifications.

Figure 7.3. *Antoine Laurent Lavoisier[12] (Paris, 1743–1794). Painted by Jacques-Louis David (1788). For a color version of this figure, see www.iste.co.uk/prudhomme/fluid3.zip*

7.4. From Lavoisier to Mendeleev

When Mendeleev (1834–1907) (see Figure 7.4(a)), a genius and visionary all-round chemist, addressed the question of a logical classification of simple bodies, many scientists had already been looking into the problem since the end of the 18th century (Bimbot 2025). In 1789, the Frenchman, Antoine Laurent Lavoisier presented a summary table of the "substances" considered at his time as chemical elements in his Elementary Treatise on Chemistry. The determination of atomic masses would later be essential for the classification of elements.

---

12 "Antoine Laurent Lavoisier, formerly Lavoisier, born August 26, 1743, in Paris and guillotined on May 8, 1794 in Paris, is a French chemist, philosopher and economist, often presented as the father of modern chemistry, which will develop from the bases and notions that he established and from a new requirement for precision offered by the instruments that he developed. He inaugurated the scientific method, both experimental and mathematical, in this field which, unlike mechanics, seemed to have to escape it". (Taken from: https://fr.wikipedia.org/wiki/Antoine_Lavoisier, where you can also read details on his conviction during the Terror in 1793).

Figure 7.4(a). *Two portraits of Dmitri Mendeleev, Russian chemist (1834–1907) author of the first Periodic Table of Elements (1869) (copyright: EU Universalis. Hulton Archive/ Getty Images)*

In 1862, the French geologist and mineralogist Alexandre-Émile Béguyer de Chancourtois was the first to note the periodicity of chemical properties as a function of the weight of atoms. He arranged the elements on a graph that he called "telluric screw" (in reference to tellurium, the central element of the graph) (see Figure 7.4(b)). This telluric screw arranged the elements in a helix on a cylinder so that those with similar properties were located one above the other; this was the first draft of the periodic classification of the elements.

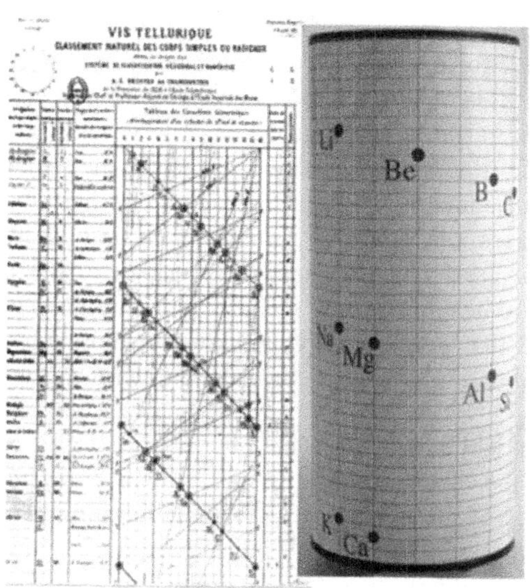

Figure 7.4(b). *The "telluric screw" of Alexandre-Émile Béguyer de Chancourtois (January 20, 1820 to November 14, 1886)*

In 1864, the British chemist John Alexander Reina Newlands flattened the fictitious cylinder carrying the helix; the elements then known are classified in eight columns of seven rows and their chemical properties are similar within each row, with sometimes two elements in the same box.

Finally, between 1864 and 1870, the German chemist Julius Lothar Meyer worked on a classification where the elements were grouped into families defined by their valence (the maximum number of bonds that an element can form depending on its electronic configuration): this was a big step toward the modern form of the periodic table; the elements were organized into groups depending on their electronic configuration, itself directly related to their valence.

In the chapter *"The Periodic Table of the Elements"*[13]: […] the establishment of the periodic table (Table 7.1) required knowledge of the chemical properties of the elements and of their atomic weight, since this classification reflects the relationship between these apparently distinct notions. While the properties of a good number of elements were relatively well known, the same was not true of their atomic weights.

The important work of chemical analysis by the Swede Jöns Jacob Berzelius (1779–1848) and the Belgian Jean-Servais Stas, in particular, led to a fairly complete and precise knowledge of the equivalent weights of the elements, i.e. the relative proportions in which they combine with each other to form a defined compound. However, to define the atomic weight, it was also necessary to know the number of atoms of each element in these compounds, i.e. its valence (this term is defined, for example, as the number of hydrogen atoms that can be fixed by an element). Avogadro's hypothesis, according to which a molecule of a compound occupies in the gaseous state the same volume as two equivalents of hydrogen, had made it possible in many cases to fix the molecular weight of the compound and, consequently, the atomic weight of the constituents and their valence. In other cases, this goal was achieved by using other properties, such as the isomorphism of crystals (same crystal structure) or the law of *Dulong* and *Petit* linking specific heats to atomic weights. An ambiguity remained, however, for certain elements that were not very well known, and the list of atomic weights and valences available to Mendeleev contained numerous errors.

Several chemists had already suspected the existence of a relationship between atomic weights and the chemical properties of elements. In 1839, Johann Döbereiner showed the existence of triads of elements with similar properties and such that the equivalent weight of one is equal to half the sum of those of the other two (such as chlorine, bromine and iodine).

13 The table is taken from https://www.gurumed.org/2016/01/05/ four-new-elements-officially-added-to-the-periodic-table.

Unlike his predecessors, Mendeleev explicitly states how his table constitutes a tool for the theoretical analysis of the properties of matter: "The chemical elements, when ordered by increasing atomic mass, show a periodicity of their chemical properties". It is thus possible to predict certain properties of the elements from their atomic mass. Mendeleev's genius is to have thought of integrating in advance several simple bodies into the table, that necessarily remain to be discovered. He therefore leaves empty boxes to be able to insert these unknown elements – which he names eka, dvi and tri, "one", "two" and "three" in Sanskrit, according to their place under an already known element, whose properties he predicts. Once found, this logic seems obvious, even if it is the result of a long reflection!

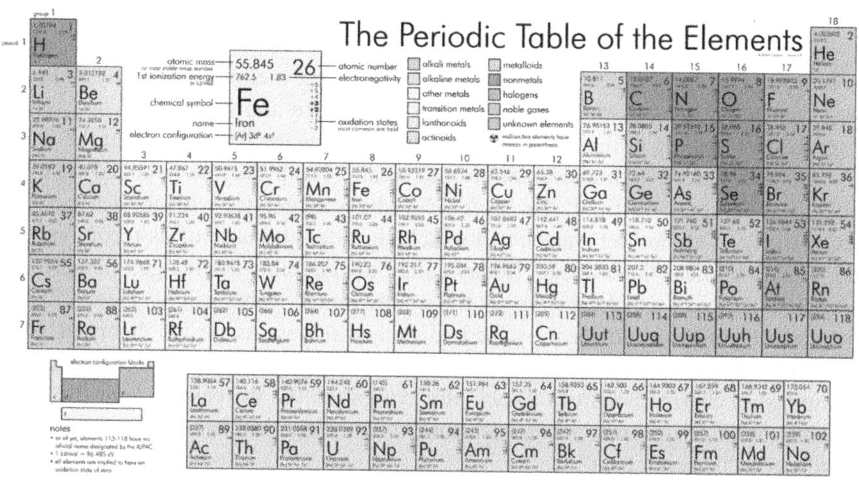

Table 7.1. *All known chemical elements (118 in number) are classified in this table which includes seven rows (or periods) and 18 columns (or groups)*[14]*. His classification, originally due to the Russian Dmitri Ivanovich Mendeleev, is based on the atomic number (number of protons), increasing along the lines, and on the distribution of electrons around the nucleus, which define the properties of the elements (each column brings together the elements having the same characteristics) (Universalis (n.d): Tableau périodique des éléments). For a color version of this table, see www.iste.co.uk/prudhomme/fluid3.zip*

A revolution did indeed take place in the field of chemistry at the end of the 18th century, signaling the end of alchemy and the theory of phlogiston. This revolution initially focused on the chemical aspect of combustion. However, what

14 The table is taken from https://www.gurumed.org/2016/01/05/four-new-elements-officially-added-to-the-periodic-table. It contains four recently discovered elements: 113, 115, 117 and 118. None of these four new elements exist in nature; they were "made" in the laboratory.

happened in relation to the other aspects? The disappearance of phlogiston and the discovery of oxygen did not immediately lead to a correct theory of combustion. Indeed, if Becher's "phlogiston" fluid had indeed disappeared as a chemical emanation of combustible bodies, Black's "caloric" fluid first appeared to designate the creation of heat and its transfer. Thermodynamics remained to be discovered, and fluid mechanics formalized. Furthermore, no one can deny the intervention of radiation in the combustion process, whether as a consequence (the radiation emitted by the flames) or as a cause (ignition by thermal radiation). Several areas of physics are thus called upon to explain combustion. We will only touch on these areas.

7.4.1. *Thermodynamics*

In 1761, Joseph Black[15] demonstrated that heating melting ice hardly raises its temperature, but simply changes the proportions of ice and water. Similarly, Black observed that heating boiling water no longer increases the temperature of the water, but only accelerates its transformation into steam. From this, he believed he could conclude that heat is a fluid, the "*caloric*", which associates with the ice particles (respectively, with boiling water), and which he called, because of its recombination with visible bodies, "latent heat" (today called the enthalpy of change of state). The caloric theory was an alternative to the controversial "phlogiston" theory, which postulated the emission of heat by the bodies themselves. As for the theory of latent heat, one of Black's main contributions to the physical sciences, it marks the beginnings of thermodynamics. Black also compared the increase in temperature of the same mass of different substances subjected to identical heating, and arrived at the notion of "specific heat".

However, it was not until the middle of the 19th century that a consistent theoretical formulation of thermodynamics appeared. Indeed, it was not until 1850 that Clausius stated[16] the two basic principles of thermodynamics. The first postulates that energy is conserved, while the second establishes the notion of the irreversibility of systems by stipulating that any transformation of a system induces an increase in entropy.

Other names to be cited among the fathers of thermodynamics are Duhem, Carnot, Onsager, and Prigogine.

15 Joseph Black (1728–1799) was a Scottish physicist and chemist. He invented a precision balance for quantitative chemical analysis that was adopted by most chemical laboratories in the late 18th century.

16 Rudolf Julius Emmanuel Clausius (1822–1888) was a Prussian physicist known for his major contributions to thermodynamics.

7.4.2. *Mechanics*

Mechanics is defined as the study of equilibrium states and movements of material systems subjected to forces. Laws, principles of mechanics; applied, experimental, theoretical mechanics.

Often seen as a branch of physics, it has theoretical components that make it appear more in the mathematical sciences. Today, it includes many specialties[17].

Until the 19th century, the concept of mechanics encompassed both the scientific study of bodies in motion and the theory of machines. A detailed history is given Wikipedia (n.d): Histoire de la mécanique.

A diagram from the French Association of Mechanics (see Figure 7.5) highlights the many areas of industry and science concerned with Mechanics.

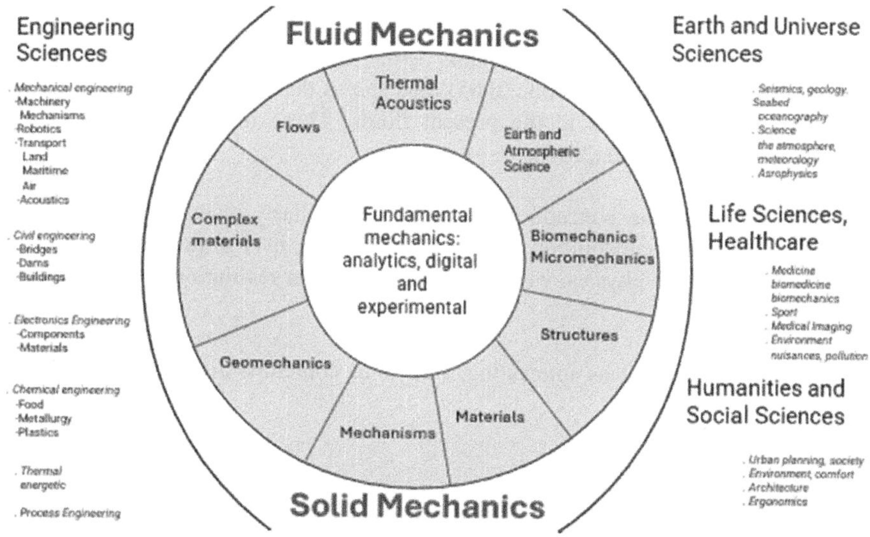

Figure 7.5. *Many fields of industry and science are concerned by Mechanics (produced by the group "University Activities in Mechanics" of the French Mechanics Association). For a color version of this figure, see www.iste.co.uk/prudhomme/fluid3.zip*

---

17 We distinguish rational mechanics, comprising the following parts: analytical, celestial, material point, solid, static, continuous media (including fluids), dynamic, quantum mechanics and relativistic mechanics.

7.5. Conclusion

The discoveries of the late 18th century cited in this chapter led to a real revolution in the conception of matter. Alchemy gave way to a new science: chemistry and considerably enriched the field of material sciences related to thermodynamics and mechanics.

The old conception of matter based on its four elements: air, fire, earth and water, has been scientifically completely outdated and replaced, on a microscopic scale, by a molecular view.

Fire, which was considered one of the building blocks of matter, simply became a reactive process between molecular elements interacting with each other.

This process is sometimes made very complex by the state of the matter (solid, liquid or gas) and its very movement.

The theory of combustion (Williams 1965; Barrère and Prud'homme 1973; Peters 2000; Poinsot and Veynante 2005) thus shows that fire zones can sometimes be considered as interfaces in the present fluids. These interfaces are then thin flames (Prud'homme 2010, 2013).

Let us note that the conception of matter, particularly reactive matter, is not limited to molecular constitution. However, this time, it is a question of nuclear chemistry and particle physics, which have brought other revolutions to the sciences of matter.

This subject, at least as interesting, but more contemporary, goes beyond the scope of this chapter.

7.6. References

Barrère, M. and Prud'homme, R. (1973). *Equations Fondamentales de l'Aérothermochimie*. Masson et Cie, Paris.

Bimbot, R. (2025). Mendeleïev Dmitri Ivanovitch (1834-1907). *Encyclopædia Universalis* [Online]. Available at: https://www.universalis.fr/encyclopedie/dimitri-ivanovitch-mendeleiev/ [Accessed 2 February 2025].

Grimaux, É. (1888). *Lavoisier : 1743–1794 d'après sa correspondance, ses manuscrits, ses papiers de famille et d'autres documents inédits*. Nabu Press, Charleston.

Peters, N. (2000). *Turbulent Combustion*. Cambridge University Press, Cambridge.

Poinsot, D. and Veynante, T. (2005). *Theoretical and Numerical Combustion*, 2nd edition. Edwards, Philadelphia, PA.

Prud'homme, R. (2010). *Flows of Reactive Fluids*. Springer, New York.

Prud'homme, R. (2013). *Flows and Chemical Reactions in Homogeneous Mixtures*. ISTE Ltd, London and John Wiley & Sons, New York.

Prud'homme, R. (2019). Notions de base sur la combustion. *Thermodynamique des interfaces et mécanique des fluides*, 19(1). doi: 10.21494/ISTE.OP.2019.0376.

Universalis (n.d). Tableau périodique des éléments [Online]. Available at: https://www.universalis.fr/media/TA161175/.

Wikipedia (n.d). Alexandre-Émile Béguyer de Chancourtois [Online]. Available at: https://en.wikipedia.org/wiki/Alexandre-%C3%89mile_B%C3%A9guyer_de_Chancourtois.

Wikipedia (n.d). Histoire de la mécanique [Online]. Available at: https://fr.wikipedia.org/wiki/Histoire_de_la_mécanique.

Williams, F.A. (1965). *Combustion Theory*, 2nd edition. Benjamin Cummings, Menlo Park.

8

Modelization of Partial Wetting by Macroscopic Approach in Discrete Mechanics

The partial wetting is generally defined by a contact angle between the liquid and the surface in the case of a static equilibrium, excluding other types of actions such as gravity, inertia, viscosity, and so on. When these last effects are no longer negligible, the modeling of two-phase flows governed by capillary forces cannot be reduced to simple geometrical laws on the surface tensions between the phases. The triple line is subject to accelerations that combine in a complex way to fix in time its motion on the surface. The macroscopic approach adopted is based on the representativity of the discrete equation of motion derived from the fundamental law of dynamics expressed in terms of accelerations. The formalism leads to a wave equation whose form corresponds to the two components of a Helmholtz–Hodge decomposition; the first to the curl-free and the second to the divergence-free. Like all other contributions, the capillary effects are expressed in two terms of the capillary potential, an energy per unit mass. The longitudinal and transverse surface tensions allow for possible anisotropy effects in the tangent plane at the interface. The assignment of the surface tension values on the triple line related to the contact angle allows us to take into account the partial wetting effects in a dynamic context. Two examples illustrate the validity of this approach.

8.1. Introduction

The physical concepts of partial wetting are now well known at the microscopic and macroscopic scales (de Gennes 1985). If there is an indisputable marker of the wettability of a surface, it is the contact angle θ_c, a reproducible measure for a liquid to wet a solid surface in the presence of a third phase, a gas mixed with the liquid's vapor. This contact angle is defined at mechanical equilibrium in the absence of any disturbance that could distort the interpretation of the measurement; gravity, an

Chapter written by Jean-Paul CALTAGIRONE and Roger PRUD'HOMME.

electric field, the geometrical structure of the surface on a small scale, the presence of impurities, which can indeed drastically modify the value of the contact angle. A general analysis by Popinet (2018) places the problem of partial wetting in the broader context of capillary flows.

Apart from the contact angle measured under ideal conditions, it is difficult to find other indisputable parameters to define the macroscopic connection of the free surface with the solid wall. Let us consider, for example, the case of a solid surface inclined by an angle α_c on which a drop of small dimension is deposited, which reaches a static mechanical equilibrium.

The observation of the latter leads us to define two angles, one upstream of the triple line in the downstream direction and the other downstream, the angles of advancing and receding. Under these conditions, the contact angle varies along the triple line defined as the trace of the liquid on the surface. It is then very difficult to exploit this result to model macroscopically and to transpose this result to similar experiments for conditions different from the initial experiment. Similarly, when there is an unsteady motion of the interface due to an external action, the contact angle θ_c is no longer respected for different reasons, (i) the viscosity of the fluid and in general its rheology; (ii) the inertia which depends in a nonlinear way on the velocity of the fluid in the vicinity of the triple line, and so on. Consequently, there are no more references allowing us to intrinsically model the motion of the triple line as a function of time. The notion of dynamic contact angle sometimes used to account for unsteady phenomena related to the triple line is not acceptable because it ignores the underlying phenomena which have an undeniable reality (Johnson et al. 1977; de Gennes 1985; Göhl et al. 2018; Zhang and Xu 2022).

Over time, the modeling associated with the contact equilibrium of several phases has been made from global formulas such as the Young–Dupré law that expresses the value of the contact angle as a function of the values of the surface tension of the liquid, vapor, and that of the solid. Even if there is a mechanical equilibrium that can be translated into a law of forces or surface tensions, the interpretation of certain quantities, for example, here the surface tension between the solid and the vapor, does not lead to an intrinsic quantification of the contact angle. Approaches based on the Cahn and Hilliard's model and a phase field method (Prud'homme 2012; Gouin and Seppecher 2017) enable us to consider the problem from a thermodynamic angle and extend the concepts to interfaces close to the critical point. The applications of wetting models are numerous, from the printed droplet microfluidics, paint spreading and undeniably justify studies of a theoretical nature (Haley and Miksis 1991; Guo et al. 2016). The methodologies used are also numerous and are based on techniques used in the more general framework of capillary two-phase flows (Guillaument et al. 2015; Legendre and Maglio 2015).

The point of view adopted in this framework is different; it postulates that a contact angle θ_c is measured under ideal conditions and adopted in the modeling. It can, like the surface tension itself, depend on the temperature but not on other physical parameters. This angle should naturally be considered as a solution of the problem if the ideal conditions are applied; in all other situations, the real contact angle will be different from θ_c.

The formulation adopted to represent the motion of a fluid in the presence of a triple line is that of discrete mechanics. This physical model postulates that the acceleration of a particle or a material medium is equal to the sum of the accelerations imposed on it by the exterior. The fundamental principle of dynamics becomes an equilibrium of accelerations, the only physical quantity considered as absolute. This model already includes the main contributions related to inertia, viscous effects, compression effects, gravitational acceleration, etc. It can be considered as an alternative to the Navier–Stokes equations for which it allows us to strictly find the same solutions, the momentum being replaced by the acceleration. The modeling of capillary effects in general and of the triple line in particular is carried out in a manner consistent with the principles of the derivation of the discrete law of motion. The objective here is to present the general model of this new approach and to analyze the role played by the contact angle in complex flows where the physical phenomena are already described.

8.2. Framework of discrete mechanics

The concepts of discrete mechanics have already been described in the literature. They can be summarized by the abandonment of some important assumptions of classical mechanics, (i) the notion of continuous medium; (ii) the notion of global reference frame $\mathbb{R}(x, y, z)$ and the change of reference frame; (iii) the derivation at a point, integration and, in general, mathematical analysis; (iv) mass and density; (v) the notion of tensor and axial or polar vectors; (vi) the quantities force and pressure; (vii) the three-dimensional view of volume; and (viii) the notion of fictitious force. These fundamental concepts of classical mechanics are replaced by those of a very different approach based on the alliance of differential geometry and electromagnetism. The global frame of reference of classical mechanics is replaced by a local frame of reference on which all effects are modeled. Unlike conventional procedures, the method we use has the particular advantage of producing results that can be used directly in numerical calculations, without the need for additional discretization.

8.2.1. *Maxwell's local frame of reference*

In the absence of an absolute reference frame, Newtonian mechanics took up the concept of velocity relativity enunciated by Galileo; this concept became the principle of Galilean relativity or the principle of inertia or Newton's first law. Any point body having a velocity in a fixed direction maintains its rectilinear motion at constant velocity if it is not subjected to an external force. Its interpretation in discrete mechanics is restricted to two parameters, (i) a length dh called discrete horizon and (ii) the time interval dt between the mechanical equilibria defined at times t^o and $t = t^o + dt$. The mechanical equilibrium corresponds to the exact satisfaction of the discrete law of motion. Length and time allow us to define a velocity or rather a celerity c such that $dh = c\, dt$. An observer located at one end of the line can only distinguish the other for a time greater than dt, hence the notion of horizon. The velocity $\boldsymbol{v}$ on this rectilinear segment cannot exceed the celerity c of the medium.

The local reference frame named Maxwell's frame of reference (Caltagirone 2024) considered is that of Figure 8.1; it is composed of the segment Γ of extremities a and b of length dh oriented by the unit vector $\mathbf{t}$. It is surrounded by a dual contour Δ oriented by the unit vector $\mathbf{n}$ such that the unit vectors are orthogonal, $\mathbf{t} \cdot \mathbf{n} = 0$. The association of several rectilinear segments connected by their extremities forms flat polygonal surfaces S bounded by a family Γ^* of segments Γ. Similarly, the contour Δ defines the dual surface $\mathcal{D}$. Two quantities are, respectively, related to the vertices of the segments and to the dual contour, the scalar potential ϕ and the vector potential $\boldsymbol{\psi}$ oriented along $\mathbf{n}$.

Space vectors do not exist in this context. The velocity $\boldsymbol{v}$ will be defined only on Γ; it is one of the components of the space velocity where knowledge is not necessary. Similarly, the acceleration γ is limited to its only component on this segment. Discrete mechanics postulates that the acceleration γ is the only absolute physical quantity, all other physical quantities being relative; the velocity is thus defined by the material derivative $\gamma = d\boldsymbol{v}/dt$ such that $\boldsymbol{v} = \boldsymbol{v}^o + \gamma\, dt$, where $\boldsymbol{v}^o$ is the so-called retarded velocity, defined at the instant t^o. Thus, without the knowledge of $\boldsymbol{v}^o$, it is not possible to predict whether the motion is, at time t, at a velocity lower or higher than the celerity of the medium, whatever the physical phenomenon of propagation considered (swell, sound, light).

The local frame of reference of Figure 8.1 represents a discrete form of the existing primal and dual structures whatever the dimension of the space considered; in one, two or three dimensions, this scheme will remain unchanged and the law of motion that follows is strictly independent of the dimension of the space. The dimension dh can take finite values related to the considered problem but it can also be reduced to zero; in this case, it is the whole frame of reference which is reduced in a homothetic way. Thus, just as the volume is reduced to zero in continuum mechanics, this approach leads to a local law. However, unlike classical mechanics, the local reference frame cannot be reduced to a point; the angles and directions are preserved. The two representations, discrete and continuous, are thus preserved.

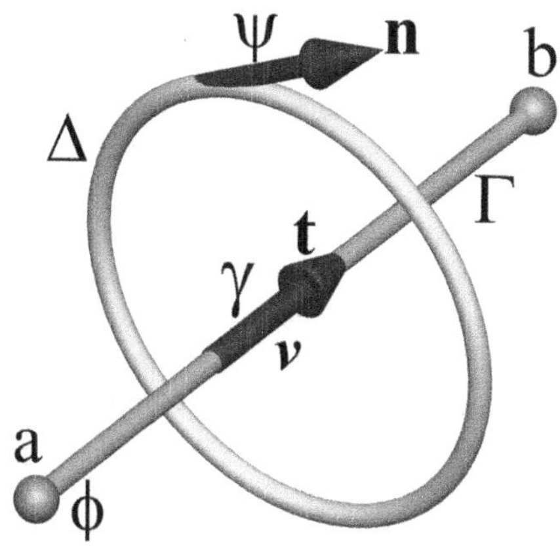

Figure 8.1. *Primal and dual geometric structures of discrete mechanics of Maxwell's frame of reference. The primal structure consists of a segment Γ oriented by the unit vector $\mathbf{t}$ with ends a and b. The dual closed contour Δ is oriented by the unit vector $\mathbf{n}$ orthogonal to $\mathbf{t}$, $\mathbf{t} \cdot \mathbf{n} = 0$. The acceleration γ and velocity v vectors are restricted to their components on Γ. The scalar potential ϕ is defined on the vertices, and the vector potential ψ is associated with the dual contour. These two structures define a direct local reference frame. For a color version of this figure, see www.iste.co.uk/prudhomme/fluid3.zip.*

The structure of the discrete reference frame evokes the remarkable idea of Maxwell (1865) to federate the concepts of magnetism and electrodynamics to define those of electromagnetism. A direct variable current created by an electric potential difference between a and b of the Γ segment induces a variable current in the Δ loop and vice versa. In direct current, the interaction disappears and the two phenomena become independent. This scheme is used in discrete mechanics to establish a law of motion in accordance with the fundamental principles of mechanics, notably Newton's second law. Since the laws of electromagnetism are currently dissociated from those of mechanics, how can the two be reconciled? This is only possible by abandoning some of the concepts of classical mechanics mentioned above. As it is not a simple analogy between two phenomena, it is necessary to find a unique law of motion applied to quantities of different nature. This transformation therefore requires the definition of common quantities by drastically reducing the number of physical quantities and the units in which they are expressed. The law of motion thus obtained will be representative of motions in fluids, of stresses and displacements in solids and of wave propagation in all media (Prud'homme 2021). The structure of the discrete equation cannot escape the more recent concepts of relativity.

Since the concepts of mathematical analysis are abandoned, they must be replaced by those of a differential geometry in the form of operators. Only four operators are needed to transform the information from the primal structure to the dual structure and vice versa. The gradient operator of the scalar potential $\nabla \phi$ is the restriction of the classical gradient to its only component on the segment Γ, $\nabla \phi = (\nabla^e \phi \cdot \mathbf{t}) \nabla^e$ where ∇^e is the gradient vector of space. The primal curl $\nabla \times \boldsymbol{v}$ is defined as the circulation of the vector $\boldsymbol{v}$ on the contour Γ^*; it is carried out by the unit vector $\mathbf{n}$. The discrete divergence $\nabla \cdot \boldsymbol{v}$ corresponds to the sum of the fluxes coming from the segments associated with the same vertex a. Finally, the dual curl $\nabla \otimes \boldsymbol{\psi}$ is set by the circulation of the vector $\boldsymbol{\psi}$ on the contour Δ and the result is assigned to the oriented primal contour Γ. The symbol $\otimes$ usually used in mechanics to denote the tensor product of two vectors is adopted because it complements in a harmonious way the concept of rotation of a pseudo-vector in classical mechanics. Since tensors do not exist in discrete mechanics, any confusion is avoided. Remarkable properties mimic those of the continuous medium:

$$\nabla \cdot (\nabla \otimes \boldsymbol{\psi}) = 0; \quad \nabla \times (\nabla \phi) = 0, \qquad [8.1]$$

since $\nabla \cdot \boldsymbol{v}$ is a scalar and $\nabla \times \boldsymbol{v}$ is a primal curl, the composition of operators provides the following identities:

$$\nabla \cdot (\nabla \otimes (\nabla \times \boldsymbol{v})) = 0; \quad \nabla \times (\nabla (\nabla \cdot \boldsymbol{v})) = 0. \qquad [8.2]$$

These properties are always verified, whatever the geometrical structure of the physical domain tessellated by the primal and dual structures; the flat surfaces formed can have any number of sides, and the volumes are any polyhedra.

Maxwell's idea leads to a very important result; the direct and induced currents are both carried by the same Γ segment, so they can add up for variable currents or flow without interacting in direct current. The representation of a vector by its three components in a global reference frame is replaced, in discrete mechanics, by vectors linked to a single direction, that of $\mathbf{t}$. The physical modeling of all direct and induced phenomena is thus to a single direction of space. The consideration of the two-dimensional or three-dimensional space is realized from cause to effect by the vertices common to two segments. The change of reference frame from $\mathbb{R}$ to a distant reference frame $\mathbb{R}'$ can only be realized under restrictive conditions; in the theory of relativity, this condition supposes that the celerity of light c_0 is constant on its whole path. In the present case, the celerity is fixed at the vertices of the primal structures and can be variable from one vertex to another and depend on other quantities, for example, temperature.

8.2.2. *Equivalence of conservation of acceleration and energy*

The modeling that leads to a unique law of motion for all wave propagation phenomena, including mechanics, requires a return to the weak equivalence principle (WEP) to modify its interpretation. Originally, this principle gives mass a privileged status; the equivalence between inertial and gravitational effects is granted to mass, "inert mass = gravitational mass". However, this principle only reflects the dynamic effects, two bodies of different natures fall with the same velocity and acceleration. Newton's second law confirms Galileo's point of view, $m\,\gamma = \mathbf{F}$ with $\mathbf{F} = m\,\mathbf{g}$.

The mass m is the same, and yet it is not removed from this relation. Special relativity will take back this notion of momentum $\mathbf{q} = m\,\boldsymbol{v}$. Even more surprisingly, the principle of equivalence between mass and energy will not allow us to remove mass from the laws of mechanics. The postulate of discrete mechanics is to discard this notion of mass and to reformulate the fundamental law of dynamics in the form:

$$\gamma = \boldsymbol{h}. \qquad [8.3]$$

The intrinsic acceleration γ of a particle or a material medium is equal to the sum $\boldsymbol{h}$ of all the accelerations applied to it. This is a local law and a law of conservation of total energy. Indeed, the integration over the length dh of the segment Γ of the acceleration is an energy per unit mass. In general, a vector can always be decomposed as a sum of a gradient of a scalar potential and a dual curl of a vector potential; it is a Helmholtz–Hodge decomposition. This decomposition is not unique a priori; it is only defined to within one $\boldsymbol{\mathcal{H}}$ field:

$$\gamma = -\nabla\phi + \nabla \otimes \boldsymbol{\psi} + \boldsymbol{\mathcal{H}}, \qquad [8.4]$$

where $\boldsymbol{\mathcal{H}}$ harmonic term, to divergence-free and curl-free.

If the scalar potential can be defined by a divergence of the velocity $\phi = \nabla \cdot \boldsymbol{v}$ and the vector potential by its curl, $\boldsymbol{\psi} = \nabla \times \boldsymbol{v}$, then the harmonic term is eliminated. The demonstration (Caltagirone 2023) indeed shows that fields with constant divergence or constant curl are discarded from the modeling of the intrinsic acceleration. They are related to uniform expansion and rotation motions with constant velocities. For example, the expansion of the Universe at constant velocity has no effect on the acceleration of a medium on a Γ segment. It follows that the acceleration, an absolute quantity, decomposes in the form $\gamma = -\nabla\phi + \nabla \otimes \boldsymbol{\psi}$. The conservation of the total energy is then written:

$$\int_\Gamma \gamma_\phi \cdot \mathbf{t}\, dl = -\int_\Gamma \nabla\phi \cdot \mathbf{t}\, dl + \int_\Gamma \nabla \otimes \boldsymbol{\psi} \cdot \mathbf{t}\, dl, \qquad [8.5]$$

where the first integral on the right side is the expansion energy and the second the rotation energy. The two accelerations $\nabla\phi$ and $\nabla \otimes \boldsymbol{\psi}$ are the only two mechanical effects to take into account. There are, however, other effects linked to gravitational, capillary, electromagnetic accelerations, etc.

The law of discrete motion describes the evolution of a system only between the times t^o and $t = t^o + dt$. It is therefore essential to define retarded potentials ϕ^o and ψ^o similar to those of electromagnetism (Liénard 1898):

$$\phi^o = -\int_0^{t^o} c_l^2 \, \nabla \cdot \boldsymbol{v} \, d\tau, \qquad \boldsymbol{\psi}^o = -\int_0^{t^o} c_t^2 \, \nabla \times \boldsymbol{v} \, d\tau. \qquad [8.6]$$

The law of motion is thus a continuous memory model; only the acceleration is absolute, and the other current quantities are defined according to their value at time t^o, the velocity $\boldsymbol{v} = \boldsymbol{v}^o + \boldsymbol{\gamma} \, dt$, the displacement $\boldsymbol{u} = \boldsymbol{u}^o + \boldsymbol{v} \, dt$, the scalar potential $\phi = \phi^o + d\,\phi$, the vector potential $\boldsymbol{\psi} = \boldsymbol{\psi}^o + d\boldsymbol{\psi}$, etc.

8.2.3. *Discrete law of motion*

The modeling of the compressive and rotational acceleration increments is done on the basis of simple experiments where the compressive energy is expressed from the divergence of the velocity $d\phi = -dt \, c_l^2 \, \nabla \cdot \boldsymbol{v}$ where c_l is the longitudinal celerity which can also be written from the isentropic compressibility coefficient χ_S. At the same time, the rotational energy is reflected by the primal curl and transverse celerity $d\boldsymbol{\psi} = -dt \, c_t \, \nabla \times \boldsymbol{v}$. The law of discrete motion then becomes:

$$\frac{\partial \boldsymbol{v}}{\partial t} = -\nabla \left(\phi^o + \frac{1}{2} |\boldsymbol{v}|^2 - c_l^2 \, dt \, \nabla \cdot \boldsymbol{v} \right) + \nabla \otimes \left(\boldsymbol{\psi}^o + \frac{1}{2} |\boldsymbol{v}|^2 \, \mathbf{n} - c_t^2 \, dt \, \nabla \times \boldsymbol{v} \right) + \boldsymbol{h}_s, \qquad [8.7]$$

where the material derivative is written as a Helmholtz–Hodge decomposition (Caltagirone 2020):

$$\boldsymbol{\gamma} = \frac{d\boldsymbol{v}}{dt} \equiv \frac{\partial \boldsymbol{v}}{\partial t} + \nabla \phi_i - \nabla^d \times \boldsymbol{\psi}_i = \frac{\partial \boldsymbol{v}}{\partial t} + \boldsymbol{\kappa}_i. \qquad [8.8]$$

The vector $\boldsymbol{\kappa}_i$ represents the curvature of the inertial potential $\phi_i = |\boldsymbol{v}|^2/2$. The two vector components are also expressed on the segment Γ and can be added or canceled according to the problem. The law of discrete motion [8.7] allows us to predict the solution of the current velocity $\boldsymbol{v}$ from its value at time t^o. The divergence and primal curl of the velocity difference, $(\boldsymbol{v} - \boldsymbol{v}^o)$, allows the potentials to be updated from the retarded potentials:

$$\text{updates} \begin{cases} \alpha_l \, \phi^o - c_l^2 \, dt \, \nabla \cdot (\boldsymbol{v} - \boldsymbol{v}^o) \longmapsto \phi^o; \\ \alpha_t \, \boldsymbol{\psi}^o - c_t^2 \, dt \, \nabla \times (\boldsymbol{v} - \boldsymbol{v}^o) \longmapsto \boldsymbol{\psi}^o. \end{cases} \qquad [8.9]$$

The quantities α_l and α_t are, respectively, the attenuation factors of longitudinal and transverse waves. When $\alpha = 1$ the corresponding energy is fully conserved and when $\alpha = 0$ the waves are fully dissipated; this is the case for a Newtonian fluid where shear leads to the dissipation of energy in the form of heat. The equation of motion being a law of conservation of total energy, the dissipated energy must be compensated in the equation; this is the role of the retarded potentials ϕ^o and $\boldsymbol{\psi}^o$. The symbol $\longmapsto$ corresponds to the actualization of these.

It is possible to group the potentials with the inertia term to obtain the Bernoulli potentials, $\phi_B^o = \phi^o + |\boldsymbol{v}|^2/2$ and $\boldsymbol{\psi}_B^o = \boldsymbol{\psi}^o + |\boldsymbol{v}|^2/2$. The equation of motion then becomes

$$\frac{\partial \boldsymbol{v}}{\partial t} = -\nabla \left(\phi_B^o - c_l^2 \, dt \, \nabla \cdot \boldsymbol{v} \right) + \nabla^d \times \left(\boldsymbol{\psi}_B^o - c_t^2 \, dt \, \nabla \times \boldsymbol{v} \right). \quad [8.10]$$

The form [8.10] shows two Lagrangians, the first for the expansion terms and the second for the rotation. Each of them is composed of a potential energy, ϕ_B^o or $\boldsymbol{\psi}_B^o$, and a kinetic energy term expressed as a function of the velocity $\boldsymbol{v}$. This form is the condition for the application of Noether's theorem (Kosmann-Schwarzbach 2011) which theoretically states that when a law can be written in the form of a Lagrangian or a Hamiltonian, then it has symmetry and invariance properties. Each of them is related to the conservation of a certain physical quantity. In this case, the time invariance ensures the conservation of the total energy, the first term of the second member conserves the compression energy and the last term ensures the conservation of the angular acceleration.

The law of motion [8.7] has other properties due to its structure. One of these shows that it is relativistic; it integrates the concepts and results of special relativity and general relativity. Consider the displacement $\boldsymbol{u} = \boldsymbol{u}^o + \boldsymbol{v} \, dt$ and suppose that the velocities c_l and c_t are equal, the classical relation of differential calculus:

$$\nabla^2 \boldsymbol{u} = \nabla \left(\nabla \cdot \boldsymbol{u} \right) - \nabla \times \nabla \times \boldsymbol{u}, \quad [8.11]$$

allows then to transform the discrete equation of motion into a nonlinear wave equation:

$$\frac{1}{c^2} \frac{d^2 \boldsymbol{u}}{dt^2} - \nabla^2 \boldsymbol{u} = -\nabla \phi^o + \nabla \times \boldsymbol{\psi}^o, \quad [8.12]$$

where the first member can be written as a Dalembertian when inertia is neglected:

$$\Box \, \boldsymbol{u} \equiv \frac{1}{c^2} \frac{\partial^2 \boldsymbol{u}}{\partial t^2} - \nabla^2 \boldsymbol{u}, \quad [8.13]$$

but here the equation is nonlinear and, by grouping the retarded terms, the equation becomes:

$$\frac{1}{c^2} \frac{d^2 \boldsymbol{u}}{dt^2} - \nabla^2 \left(\boldsymbol{u} - \boldsymbol{u}^o \right) = 0. \quad [8.14]$$

It is then easy to show that a wave equation is relativistic. It is enough to apply the Lorentz transformation of the reference frame $\mathbb{R}$ to show an identical equation corresponding to the reference frame $\mathbb{R}'$. It thus remains invariant with respect to the considered transformation. This invariance is not incompatible with the invariance resulting from the Galilean transformation related to inertia. On the contrary, it is necessary to (i) ensure the filtering of uniform translational and rotational motions and (ii) describe the blocking of the velocity of a particle in its accelerated motion on a linear trajectory at the celerity of the medium.

The nonlinear wave equation [8.14] is naturally deduced from Newton's second law. It is not derived from a Lorentz transformation nor from the use of the factor $\gamma = 1/\sqrt{1 - v^2/c^2}$. The approaches of SRT and discrete mechanics are very different, even if the results are identical. The discrete model presented here allows us to fully recover the solutions of the Navier–Stokes equation for fluids in a wider context describing waves in general (swell, acoustic waves, light).

The boundary conditions associated with the law of motion [8.7] are directly integrated in the latter in the form of operators, $-\nabla q + \nabla \otimes \boldsymbol{r}$ where q corresponds to a surface flux and $\boldsymbol{r}$ to a shear stress at the walls. These two potentials allow us to simulate any kind of boundary condition in a way consistent with the formulation.

8.3. Concepts of capillary motion

The abandonment of certain concepts of classical mechanics has an impact on the treatment of capillary flows. First of all, the discarding of mass and density has an effect on the quantities used to simulate two-phase flows in the presence of capillary effects. The surface tension γ used classically becomes $\sigma = \gamma/\rho$, named surface tension per unit mass; this one is expressed in $m^3 \ s^{-2}$. Its product by the curvature κ is an energy per unit mass, the capillary energy $\sigma \ \kappa$ and, like the others, it will be a scalar or vector potential depending on the term in which it appears. The experimental values assigned to γ will also be those used for σ because the pressure itself becomes $\phi^o = p/\rho$. The density thus disappears from all of the adimensional quantities and parameters that usually characterize capillary effects. Moreover, we find the grouping γ/ρ in most of them.

8.3.1. *Capillary motion law*

The capillary potentials noted $\phi_c = \sigma \ \kappa$ and $\boldsymbol{\psi}_c = \sigma \ \kappa \ \mathbf{n}$ are also energies per unit mass, and the quantity $c_c = \sqrt{\sigma \ \kappa}$ is the capillary wave celerity which can be interpreted as the capillary velocity $\boldsymbol{v}_c$. The capillary acceleration γ_c can also be modeled by stating that it is equal to the sum of the accelerations imposed on it. This leads to an equation of capillary motion:

$$\begin{cases} \gamma_c = \dfrac{d\boldsymbol{v}_c}{dt} = -\nabla \left(\phi_c^o - \sigma_l \, \kappa_l \, dt \, \nabla \cdot \boldsymbol{v}_c \right) + \nabla \otimes \left(\boldsymbol{\psi}_c^o - \sigma_t \, \kappa_t \, dt \, \nabla \times \boldsymbol{v}_c \right), \\ \phi_c^o - \sigma_l \, \kappa_l \, dt \, \nabla \cdot \left(\boldsymbol{v}_c - \boldsymbol{v}^o \right) \longmapsto \phi_c^o, \\ \boldsymbol{\psi}_c^o - \sigma_t \, \kappa_t \, dt \, \nabla \times \left(\boldsymbol{v}_c - \boldsymbol{v}^o \right) \longmapsto \boldsymbol{\psi}_c^o. \end{cases} \quad [8.15]$$

It has exactly the same structure as the equation of motion. It is of course free of the compression and rotation terms of the general law. It applies to the case of a Σ surface, for example, a liquid sheet having a surface tension σ and a local curvature κ, these two quantities being defined for the longitudinal and transverse waves. This equation [8.15] allows us to calculate the evolution of the velocity over time of a surface subjected to motions generated by variations of surface tension and curvature. It is possible to represent the stationary solution of a catenoid surface, a minimal surface constrained by two circles, from any initial solution.

Since capillary effects are sometimes modified by gravity, it is desirable to rewrite the equation of fluid dynamics subjected to this acceleration; this should of course lead to the stationary solutions of fluid statics, compressible or not. Its modeling is similar to the general law of motion where the gravitational energy is equal to $g\,r$, where r is a fixed coordinate with respect to a reference frame; it reads:

$$\begin{cases} \gamma_g = \dfrac{d\boldsymbol{v}_g}{dt} = -\nabla \left(\phi_g^o - \mathbf{g} \cdot \mathbf{t} \, r \, dt \, \nabla \cdot \boldsymbol{v}_g \right) + \nabla \otimes \left(\boldsymbol{\psi}_g^o - \mathbf{g} \cdot \mathbf{n} \, r \, dt \, \nabla \times \boldsymbol{v}_g \right), \\ \phi_g^o - \mathbf{g} \cdot \mathbf{t} \, r \, dt \, \nabla \cdot \left(\boldsymbol{v}_g - \boldsymbol{v}^o \right) \longmapsto \phi_g^o, \\ \boldsymbol{\psi}_g^o - \mathbf{g} \cdot \mathbf{n} \, r \, dt \, \nabla \times \left(\boldsymbol{v}_g - \boldsymbol{v}^o \right) \longmapsto \boldsymbol{\psi}_g^o, \end{cases} \quad [8.16]$$

where $\nabla(\mathbf{g} \cdot \mathbf{t}\, r) = (\mathbf{g} \cdot \mathbf{t})\, \mathbf{t}$ corresponds to the projection of the gravity vector onto Γ; the other component projects in the same way onto Δ.

The equation of motion of capillary effects [8.15] and that of fluid statics [8.16] can be combined in the general law of motion [8.7]. The new scalar potential is then written $\phi^o + \phi_c + \phi_g \longmapsto \phi^o$ and similarly for $\boldsymbol{\psi}^o$ and for the global velocity $\boldsymbol{v}$.

Since all potential source terms are written in the form of a Helmholtz–Hodge decomposition, the form of the general equation remains unchanged:

$$\begin{cases} \dfrac{d\boldsymbol{v}}{dt} = -\nabla \left(\phi^o - \left(c_l^2 + \sigma_l\, \kappa_l + g\, r\right)\, dt\, \nabla \cdot \boldsymbol{v}\right) \\ \qquad +\nabla \otimes \left(\boldsymbol{\psi}^o - \left(\dfrac{\nu}{dt} + \sigma_t\, \kappa_t + g\, r\right) dt\, \nabla \times \boldsymbol{v}\right), \\ \phi^o - \left(c_l^2 + \sigma_l\, \kappa_l + g\, r\right)\, dt\, \nabla \cdot (\boldsymbol{v} - \boldsymbol{v}^o) \longmapsto \phi^o, \\ \boldsymbol{\psi}^o - \left(\nu/dt + \sigma_t\, \kappa_t + g\, r\right)\, dt\, \nabla \times (\boldsymbol{v} - \boldsymbol{v}^o) \longmapsto \boldsymbol{\psi}^o, \end{cases} \qquad [8.17]$$

where ν is the kinematic viscosity that replaces the generic term $dt\, c_t^2$ for viscous fluids with large time constants where these effects are always present.

The groupings $dt\, \nabla \cdot \boldsymbol{v}$ and $dt\, \nabla \times \boldsymbol{v}$ are dimensionless; the explicit presence of the time lapse dt between two mechanical equilibria has a very special meaning that does not exist for the Navier–Stokes equations. For small time constants, the exchanges between the potential energy, fixed by ϕ^o or $\boldsymbol{\psi}^o$, and the kinetic energy, associated with $\nabla \cdot \boldsymbol{v}$ or $\nabla \times \boldsymbol{v}$, are important. For large time constants, the products $dt\, \nabla \cdot \boldsymbol{v}$ and $dt\, \nabla \times \boldsymbol{v}$ remain of the same order of magnitude as the other terms in the equation, but $\nabla \cdot \boldsymbol{v}$ and $\nabla \times \boldsymbol{v}$ tend to zero or to constants. The acceleration $\boldsymbol{\gamma}$ becomes zero when the motion is stationary and the two terms $-\nabla \phi + \nabla \otimes \boldsymbol{\psi}$ define the new mechanical balance.

Let us now consider the respective roles of each of the expansion c_l^2, capillary $\sigma\, \kappa_l$ and gravitational $g\, r$ and those of the vector potentials. Let us fix for that orders of magnitude compatible with the capillary movements in a viscous liquid subjected to gravity, $\sigma_l = 10^{-4}\, m^3 s^{-2}$, $\kappa = 1/R = 10^3\, m$, $c_l = 10^3\, ms^{-1}$, $g = 10\, ms^{-2}$ and $\nu = 10^{-6}\, m^2 s^{-1}$. Let us summarize the different potential groupings by considering that is of the same order of magnitude as the ratio of a dimension by a celerity $dt = 1/(\kappa\, \sqrt{\sigma\, \kappa}) \approx 10^{-2}\, s$:

$$\frac{\sigma_l\, \kappa_l}{c_l^2} = 10^{-7}; \quad \frac{g}{\kappa_l\, c_l^2} = 10^{-8}; \quad \frac{\nu}{dt\, \sigma_t\, \kappa_t} = 10^{-3}; \quad \frac{\nu\, \kappa_t}{dt\, g} = 10^{-2};$$

$$\frac{g}{\sigma_l\, \kappa_l^2} = 10^{-1}; \quad \frac{g}{\sigma_t\, \kappa_t^2} = 10^{-1}. \qquad [8.18]$$

The groupings $(g/\kappa_l)/c_l^2$ and $(g/\kappa_l)/c_l^2$ show that the effects due to capillarity or gravity are, respectively, seven and eight orders of magnitude smaller than those related to the compressibility of the liquid and can therefore be neglected. The same is not true for rotation where the quantities related to viscosity, capillarity and gravity form sets closer to unity. Of course, this exercise must be reproduced according to the problem at hand, but it shows the interest of such a global approach to define adimensional parameters whose quantities are all expressed exclusively by a length and a time. Contrary to the classical adimensional analysis, the adimensional parameters are not defined by orders of magnitude of the velocity, the density or a characteristic length whose choices can be discussed.

There are many dimensionless parameters related to capillary effects, for example, the Weber number $We = \rho\, v^2\, L/\gamma$, the capillary number, $Ca = v\, \mu/\gamma$, the Froude number, $Fr = v/\sqrt{g\, L}$, the Laplace number, $La = \gamma\, L\, \rho/\mu^2$, the Ohnessorge number, $Oh = \mu/\sqrt{\rho\, \gamma\, L}$ or the Bond number, $Bo = \rho\, g\, L^2/\gamma$, to name a few. They are not all independent and can be combined to form others. There are of course a few in the list [8.18] including the Bond number $Bo = g/(\sigma\, \kappa^2)$ as the ratio of gravitational energy to capillary energy.

The most notable difference is the presence of the ratio ν/dt, representing the energy per unit mass of viscous friction. The true notion of viscosity is complex. The fact that it is measured by steady-state experiments does not give it the desired general character. Moreover, Newton's shear law $\tau = \mu\, dv/dy$ induces a paradox in the Navier–Stokes equation at small time constants when a sudden shear is imposed on a fluid at a given time. The perturbation is theoretically instantaneously felt at infinity, and the initial stress is itself infinite. The hyperbolic character of the discrete equation [8.7] overcomes this drawback.

8.3.2. *Modelization of partial wetting*

This section presents how this formulation [8.17] is likely to take into account the notion of partial wetting between a surface and a liquid. Many models are based on the static contact angle which is sometimes extended in dynamics. In fact, these approaches are set aside in the framework of the presented formulation. Indeed, it is very difficult to predict what happens to the contact angle when capillary effects are intermingled with those due to viscosity, inertia, gravity, etc. The principle adopted here is different; the modeling of all of the effects is carried out in a coherent way and, if the equation of motion is representative of the phenomena, it is then used without trying to analyze each phenomenon separately. Indeed, the instantaneous local angle is a complex function of the terms of the equation of motion. Young–Dupré's law, $\cos\theta = (\gamma_{SV} - \gamma_{SL})/\gamma_{LV}$ where γ_{SV} is the surface tension between the solid and the vapor, γ_{SL}, the surface tension between solid and liquid and γ_{LV} the surface tension between liquid and vapor, provides an equilibrium relation but does not allow us to describe the dynamics of the interface formation; moreover, the tension γ_{SV} is difficult to quantify.

In the case of capillary flows without connection to a triple line, the equation [8.17] allows us to find the classical results (Caltagirone 2021). For example, the solution of the Laplace problem of a drop of average curvature $\kappa_l + \kappa_t = 2/R, p/\rho = \phi_c = 2\,\sigma\,\kappa$

is obtained exactly (to machine precision). The dynamic case of an ellipse oscillating around a mean position is solved precisely (to second order) from an arbitrary Lagrangian–Eulerian method to obtain the theoretical oscillation frequency equal to $f = m\,(m^2 - 1)\,\sigma\,R^3$ (Prosperetti and Tryggvason 2007). Many other cases of two-phase flows show the validity of the proposed approach. In all cases, the static or dynamic solution of the problem does not depend on the density, which is integrated both in the capillary potential $\phi = p/\rho$ and in the surface tension per unit mass, $\sigma = \gamma/\rho$. Moreover, the notion of average curvature is replaced by the two curvatures κ_l and κ_t which form an orthogonal reference of the tangent plane of the interface.

Let us now introduce the concept of partial wetting in the discrete model associated with the equation of motion [8.17]. In fact, this equation remains unchanged; it is only necessary to introduce an information on the desired contact angle at the capillary equilibrium. Indeed, this notion of contact angle defined as a property of a surface–liquid couple is only realized in one case, the one where all the other effects, inertial, viscous, gravitational, compressible are negligible before the surface tension. In this situation, the shape of the surface of the liquid in contact with the solid is a portion of a sphere and this is also true for more complex situations, for example, a lens of a liquid in equilibrium between two other liquids. In other cases, the contact angle is not respected either in static or in dynamic. Simplistic laws or reasoning neglecting other effects cannot represent the reality of such complex phenomena. The point of view adopted is based on an equation of motion which already correctly describes all physical phenomena. Only a global model allows us to assemble the contributions linked to each of them in an essentially macroscopic vision; the effects at very small scale will not be taken into account.

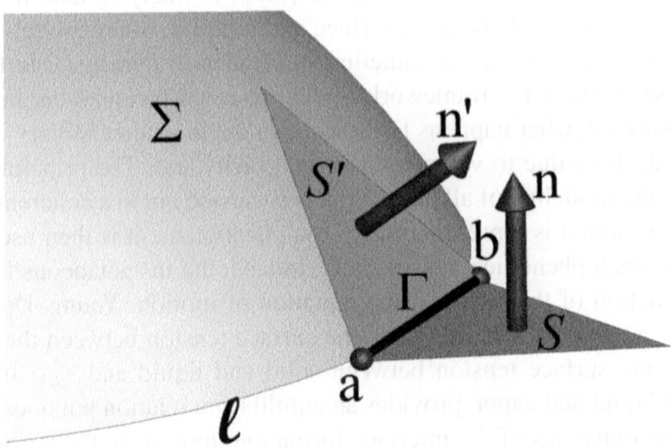

Figure 8.2. *Drop of surface Σ placed on a solid plane; two facets S and S', associated with the common segment Γ of the triple line ℓ, respectively, of normals $\mathbf{n}$ and $\mathbf{n}'$ define the dihedral angle additional to the contact angle θ_c. For a color version of this figure, see www.iste.co.uk/prudhomme/fluid3.zip.*

Let us consider the Σ interface of Figure 8.2, for example, that of a drop of liquid on a solid surface partially wetted by it. The liquid is in contact with the solid on a surface delimited by the triple line $\boldsymbol{\ell}$. This line is formed by the line segments Γ whose ends a and b are oriented by the unit vector t. The interface Σ and the solid surface are tessellated by surfaces such as $\mathcal{S}$ and $\mathcal{S}'$ which have, in this case, the triple line segment $\boldsymbol{\ell}$ in common. The normals to these two facets are called $\mathbf{n}$ and $\mathbf{n}'$. As in the general case of capillary flows, the capillary accelerations are written as a Helmholtz–Hodge decomposition:

$$-\nabla \left(\sigma_l \, \kappa_l\right) + \nabla \otimes \left(\sigma_t \, \kappa_t \, \mathbf{n}\right). \qquad [8.19]$$

The first term is evaluated from the quantity $\sigma_l \, \kappa_l$ defined at the vertices of the segment Γ, and the second term is the dual curl related to the circumferences of the two primal facets. The center of $\mathcal{S}$ located at c is the intersection of the perpendicular bisectors of the triangle sides. In general, the curvature is related to the divergence of the normals on the interface. The use of the dihedral angle is particularly well suited in this context because the normals $\mathbf{n}$ and $\mathbf{n}'$ are defined exactly from the coordinates of the vertices of each triangle. The dihedral angle θ_d is defined by the two normals $\mathbf{n}$ and $\mathbf{n}'$ such that $\cos \theta_d = \mathbf{n} \cdot \mathbf{n}'$; it is the supplementary angle of the contact angle $\theta_c = \pi - \theta_d$. It is possible to define a contact curvature named $\boldsymbol{\kappa_\ell}$ by the outer product of the normals $\mathbf{n}$ and $\mathbf{n}'$ attributed to the segment Γ oriented by $\mathbf{t}$ which is obtained by considering a distance, the one joining the circumferences of the two triangles:

$$\boldsymbol{\kappa_\ell} = \frac{\mathbf{n} \wedge \mathbf{n}'}{\|c \, c'\|}, \qquad [8.20]$$

where $\|c \; c'\|$ is the norm of the distance between the circumcenters of the two facets $\mathcal{S}$ and $\mathcal{S}'$. The symbol $\wedge$ defines the exterior product which corresponds to the component normal to the considered surface of the vector product noted with the symbol $\times$. In the absence of a global reference frame, the outer product is particularly adapted to the writing of the laws of discrete mechanics. Unlike the vector product which is defined for a three-dimensional space, the outer product is defined for any n dimensional vector space.

The quantity $\boldsymbol{\kappa_\ell}$ can also be named curvature of the triple line; it characterizes indeed, with the surface tension $\sigma_{\boldsymbol{\ell}}$, the acceleration due to the partial wetting. Thus, the transverse curvature κ_t applied to all the facets of the interface Σ will take, on the triple line, the value $\boldsymbol{\kappa_\ell}$. The global formulation then remains unchanged; only the transverse curvatures will be kept equal to $\boldsymbol{\kappa_\ell}$ during a simulation. The link between the experimentally evaluated contact angle θ_c and the curvature of the triple line is direct. In the case of a drop of radius R, the contact curvature, at capillary equilibrium, is equal to $\boldsymbol{\kappa_\ell} = 2/R$. From a practical point of view, the calculation of the curvature of a tessellated interface by triangles computed from the formula [8.20] provides a good approximation of the theoretical value.

Let us consider the case of an evolution in time of the shape of an interface Σ subjected to various accelerations due to the effects of gravity, viscosity, inertia, etc. The value of the contact curvature κ_ℓ, once defined as a function of the contact angle θ_c, is maintained over time on the triple line of the equation [8.17] but not, for all that, the value of the normals $\mathbf{n}$ and $\mathbf{n}'$; and, in general, the shape of the interface is not fixed. It appears only as a constraint to be satisfied, in the absence of these actions, to reach the capillary equilibrium. For example, let us consider a drop that spreads on a partially wetting surface characterized by a contact angle θ_c in the presence of a gravitational field. In this case, the contact angle at capillary equilibrium is not θ_c; it results from an equilibrium, including the gravitational acceleration. The velocity at equilibrium is strictly zero.

Thus, mechanical equilibrium, whether static or dynamic, cannot be defined by a single parameter such as the contact angle (static or dynamic). The physics of these phenomena is complex but it is inscribed in the law of motion [8.17]. Its resolution is essentially computational because it is nonlinear and the interactions between the different phenomena it models are particularly intertwined. From an initial condition fixed by the shape of the interface Σ and the source terms at time t^o, the solution of the equation allows us to predict the solution on the velocity $\boldsymbol{v}$ at the current time $t^o + dt$; the potentials ϕ^o and ψ^o are updated, and the new shape of the interface is computed using this velocity. If a static equilibrium is possible, the velocity tends to zero, but all terms that no longer depend on it must still strictly satisfy the equation. A conservation equation must always be satisfied, even when the velocity tends to zero; no residual velocities must remain, for example, the unphysical stray currents frequently reported in the literature.

8.4. Analyses of some capillary flows

8.4.1. *Spreading and equilibrium of a drop*

The dihedral angle and the contact curvature κ_ℓ are implemented in the case of the spreading of a drop of liquid partially wetting the substrate on which it is placed under the assumption that the initial equilibrium is not assured. In this situation, the liquid spreads or retracts in order to finally satisfy the theoretical contact angle θ_c between the liquid and the solid surfaces.

A drop, having the shape of a semicircle of radius $R = 0.625\ 10^{-3}\ m$, is placed on a plane whose wettability can be modified over time from the contact curvature κ_ℓ. The density of the fluid is equal to $\rho = 1000$, and its viscosity is equal to $\mu = 10^{-2}\, kg m^{-1} s^{-1}$ and the external medium is air. The curvature of the initial semicircle is equal to $\kappa_\ell = 1600$. From the initial time, we impose a contact curvature $\kappa_\ell = 750$ to simulate a wetting surface and, from a time equal to $t = 0.2\, s$, we modify this value which is brought to $\kappa_\ell = 2230$ to represent a non-wetting surface. Table 8.1 gives the characteristics of the drop for each step of the simulation.

Curvature	Radius of meniscus	Contact angle	Capillary pressure
$1600\ m^{-1}$	$0.625\ 10^{-3}\ m$	$90°$	$112\ Pa$
$750\ m^{-1}$	$1.333\ 10^{-3}\ m$	$48.2°$	$52.5\ Pa$
$2230\ m^{-1}$	$0.415\ 10^{-3}\ m$	$150°$	$155.04\ Pa$

Table 8.1. *Evolution of a drop on a plane; characteristics of the meniscus for each step of the simulation*

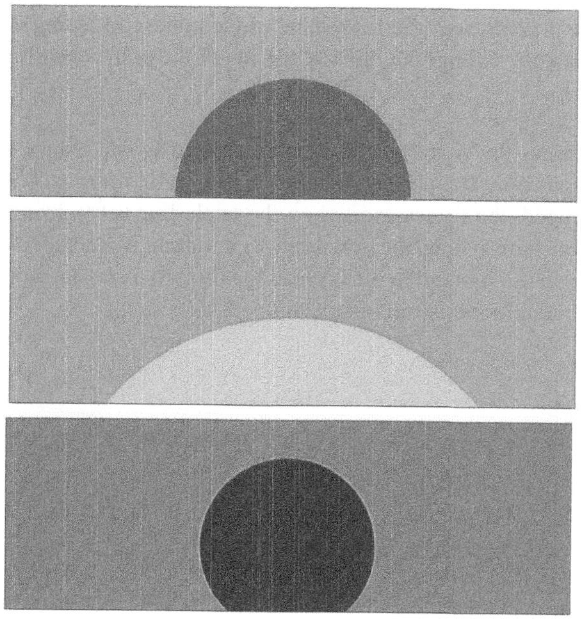

Figure 8.3. *Simulation of the evolution of a drop on a flat surface as a function of the curvature applied to the wall. On top, a drop in equilibrium where the curvature* $\kappa_\ell = 1600$ *and the contact angle* $\theta_c = 90°$ *are given as initial condition; in the center, the curvature* $\kappa_\ell = 750$ $(\theta_c = 48.2°)$ *is imposed on the triple line and at a time of* $t = 0.2\ s$*, we impose* $\kappa_\ell = 2230$ $(\theta_c = 150°)$*. For a color version of this figure, see www.iste.co.uk/prudhomme/fluid3.zip*

The simulation is performed using equation [8.17] and maintaining on the contact line the κ_ℓ curvature in two space dimensions. The computational methodology used is that described previously (Caltagirone 2021); the unstructured primal mesh, composed of regular triangles, tessellates the two domains, the initially hemispherical drop and the complement of the rectangular cavity shown in Figure 8.3. The interface tracking method is similar to front-tracking (Prosperetti and Tryggvason 2007); the triple line corresponds to a line of the mesh, and this and all vertices of the primal mesh are advected by the velocity field $\boldsymbol{v}$. The boundary condition on the horizontal plane corresponds to a non-slip except on the segment where the junction point of the Σ interface and the plane is located, i.e. on the triple line. Indeed, the condition of adhesion would lead to a zero velocity of the triple line on the plane which is false. In fact, this problem of inadequacy of the boundary condition of the triple line results from the globalization of the velocity field $\boldsymbol{v}$ which contains in fact two components, one at the absence of divergence and the other at the absence of curvature. The triple line must be free to move on a solid surface, which corresponds only to the condition $\boldsymbol{v} \cdot \mathbf{n} = 0$. In the present case, the Γ segment of the plane containing the vertex of the triple line is thus kept sliding, the other segments of the plane ensuring the condition of no sliding.

Figure 8.4 shows the shape of the drop in each quasi-stationary state where the velocities are negligible. The pressure inside the drop is then almost constant. As soon as the κ_ℓ is changed, the forces exerted on the triple line act to bring the drop back to an equilibrium state where the κ_ℓ contact curvature is satisfied. Over time, the velocities evolve in the domain first to spread the drop to a contact angle of 48.2° and then to contract the interface to a contact angle of 150°.

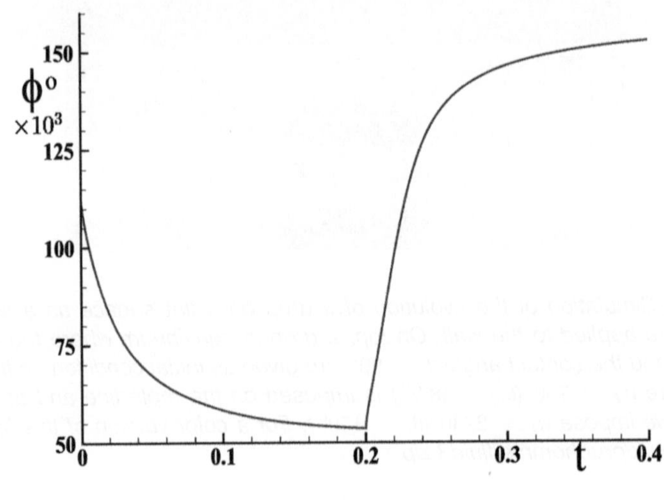

Figure 8.4. *Evolution of the average energy difference per unit mass (in $m^2\ s^{-2}$) between the drop and the exterior on a flat surface as a function of time with respect to curvature variations at the wall. The asymptotes of the capillary potential lie at $\phi^o = 52.5\ 10^{-3}$ and $\phi^o = 155\ 10^{-3}$. For a color version of this figure, see www.iste.co.uk/prudhomme/fluid3.zip*

Figure 8.4 shows the evolution of the scalar potential difference ϕ^o between the interior and exterior of the drop as a function of time; in each of the two domains, liquid and gas, the potential is uniform and the velocity and inertia become very low very quickly after a phase of rapid evolution of the interface. The results obtained from the computationally simulated data are consistent with the theoretical predictions. In this example, the interfaces correspond exactly to a constant curvature, for example, of spheres (or cylinders in two spatial dimensions). This is of course no longer the case when gravity is taken into account; the interfaces between air and liquid are no longer spheres and the contact angle θ_c is no longer respected; the real angle of static equilibrium is a complex function of the capillary potentials ϕ_c and that of gravity ϕ_g.

This first characteristic example of capillary flows translates the principle of the use of the law of discrete motion (equation [8.17]); only the physical parameters are adjusted according to the considered problem but the form of the equation itself remains unchanged. For example, when the incompressibility of the flow is required, it is not the divergence term of the velocity $\nabla \cdot \boldsymbol{v} = 0$ that is removed from the equation but the value of the celerity is increased. Similarly, when a contact angle θ_c between a liquid and a solid is desired, it is not maintained explicitly and will not be assured if there are other constraints.

8.4.2. *Capillary rise between two vertical planes*

The problem of the capillary rise of a liquid in a pipe is particularly emblematic of capillary flows. The phenomenon is simple to describe; the liquid rises in the pipe of radius r and partially wets the walls of the pipe, eventually stabilizing at a height h fixed by Jurin's law. This law is established on the basis of an equilibrium taking into account the weight of the liquid column of density ρ, $F_p = \pi\, r^2\, h\, \rho\, g$ and that of the capillary force given by $F_c = 2\,\pi r\, \gamma\, \cos\theta_c$ where θ_c is the contact angle and γ is the surface tension; an energy balance per unit mass instead of a force balance leads to the same conclusion:

$$h = \frac{2\,\gamma\, \cos\theta_c}{\rho\, g\, r} = \frac{2\,\gamma}{\rho\, g\, R} = \frac{\sigma\, \kappa}{g}, \qquad [8.21]$$

where $\sigma = \gamma/\rho$; as for the other laws related to capillary effects, it is the surface tension that appears naturally.

The case proposed here is that of a liquid of density $\rho_2 = 1000$ and viscosity μ_2 in capillary ascent between two planes whose contact angle θ_c with the vertical walls corresponds to $\cos\theta_c = 0.9$; the average curvature of the interface is thus equal to $\kappa_\ell = 1800$. The medium above the liquid has density $\rho_1 = 1$ and viscosity μ_1. The surface tension $\gamma = 0.055555$ is constant. The domain is planar, rectangular, of dimensions $[-0.0005, 0.0005] \times [0, 0.01]$, and the base of the interface between the two fluids in static equilibrium is placed at $y = 0.005$. The value of gravity in the downward vertical direction is chosen to obtain an upward or downward shift of the interface or static equilibrium. The primal topology corresponds to a mesh based on regular triangles bounding the two media by a mesh line. The boundary conditions are associated with vertical solid walls ($\mathbf{v} = 0$) and horizontal free walls where the pressure is constant and equal to $p = 0$. The pressure difference on both sides of the interface resulting from the simulations must be equal to $\Delta p = \gamma\, \kappa$ or $\phi = \sigma\, \kappa$ at static equilibrium.

When gravity is less than the value corresponding to static equilibrium, capillary forces are dominant and the heavier fluid is pulled upward near the walls. These same forces then act to restore a shape close to the circle and a curvature equal to $\kappa_\ell = 1/R$ where R is the radius of curvature. The unsteady vortex motions are induced by all the physical phenomena present in the problem. On the contrary, when gravity is more important, the associated forces pull the interface down until static equilibrium is reached in more or less important times which depend on the exchanges between inertia and viscosity forces. We observe that the boundary conditions of the problem are well satisfied, in particular that the top and bottom of the cavity are at the same pressure and that the pressure jump at the interface corresponds to the exact value $\Delta p = 100$. The value of gravity used for the static situation is equal to $g = 19.636112$. We note that the height of the meniscus h to be considered to find Jurin's law is not the minimum height of the meniscus on the axis but the average value which is $h = 5.092657\, 10^{-3} m$.

Figure 8.5 shows the result obtained for the static equilibrium; the velocity is zero in the whole domain at the accuracy of the machine, and the two potential fields of compression ϕ^o and rotation $\boldsymbol{\psi}^o$ alone define this mechanical equilibrium:

$$-\nabla \phi^o + \nabla \otimes \boldsymbol{\psi}^o = 0. \qquad [8.22]$$

The two scalar and vector potentials integrate, respectively, for compression and rotation, the energies accumulated over time by the terms of viscosity, inertia, gravity and capillary effects. The law [8.22] represents what remains of the equation of motion [8.17] when the velocity is equal to $\mathbf{v} = 0$. Thus, the dynamics of the capillary rise leads to a static state which respects Jurin's law; this is what is observed experimentally; the velocity becomes strictly zero. This result is remarkable from the point of view of mechanical equilibrium in a system where the phenomena are complex and intertwined. From a mathematical point of view, the relation [8.22] translates the equality of two orthogonal components which are both projected on the same segment Γ and which can thus add up. This simulation, carried out from the Navier–Stokes equation, with strictly the same parameters, does not lead at all to the same conclusions; when the velocity is equal to zero in the latter, we obtain $\nabla p^o = 0$ or $p^o = 0$, a wrong solution. Since static equilibrium cannot be reached, the velocities predicted by the simulation can never be attenuated and large spurious currents remain near the triple line. This result is one of the paradoxes of the Navier–Stokes equation, which does have the equivalent of the scalar potential, the pressure, but does not have the vector potential that accumulates the shear stress.

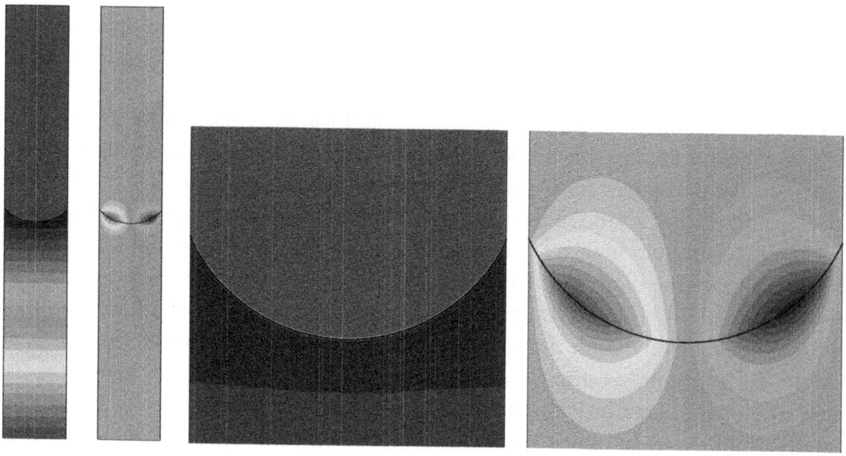

Figure 8.5. *Static mechanical equilibrium defined by the two scalar ϕ^o (left) and vector ψ^o (right) potentials; the two images on the right show the details of these two potentials near the interface. For a color version of this figure, see www.iste.co.uk/prudhomme/fluid3.zip*

Static equilibrium [8.22] represents a general law satisfied in particular by an elastic solid subjected to compressive and shear stresses as well as in the case of a fluid whose velocity physically tends toward zero, as in the example treated in this section. Moreover, it is valid in the case of a stationary flow where the velocity field $\boldsymbol{v}$ does not depend on time but remains defined by all effects, viscous, gravitational, inertial, capillary, etc. There are two possibilities, (i) to use only one of the potentials, ϕ^o or ψ^o, and obtain a non-zero velocity field $\boldsymbol{v}$; and (ii) to keep both potentials which leads to a zero velocity field. In the first case, the solution is written $(\phi^o, \boldsymbol{v})$ or $(\psi^o, \boldsymbol{v})$ and in the second case, it takes the form (ϕ^o, ψ^o); the velocity field is not part of the solution anymore. We show that the knowledge of one of the potentials introduced in the discrete equation of motion leads instantaneously to the velocity field $\boldsymbol{v}$ retained previously. In the case of persistent unsteady solutions, for example, for a periodic, chaotic or turbulent flow, the acceleration is non-zero and all of the terms of the discrete equation are nested.

8.5. Conclusions

The modeling of two-phase flows dominated by capillary phenomena requires us to retain multiple other physical phenomena related to viscosity, gravity, inertia, etc.

It is difficult, if not impossible, to predict the evolution of these flows over time or even the static solutions except in a few canonical examples. The principle adopted in discrete mechanics is to derive an equation of motion as general as possible, likely to represent at least the flows of fluids, the stresses and displacements of solids or the propagations of waves. The law of motion thus established requires a thorough validation that will give it sufficient legitimacy to apprehend other complex situations.

The law of motion, which can already account for two-phase flows in general, is confronted with capillary phenomena related to the triple line. The contact angle is most often the parameter most used to translate the phenomenon of partial wetting between a liquid and a solid in the presence of a gas. This parameter is well measured by different techniques in the case of a static equilibrium. As the spatial scales chosen in the presence of the Earth's gravity are necessarily small, the time constants are small and the static equilibrium is reached very quickly. The application of an essentially static parameter to dynamic flows must be done with some precautions; it is not possible, for example, to impose a contact angle, static or dynamic, explicitly. The notion of contact curvature is more flexible because it does not have to be satisfied formally; it allows us to obtain the correct contact angle at static equilibrium without explicitly imposing its value in dynamics. The two examples provided highlight some aspects of its use.

8.6. References

Caltagirone, J.-P. (2020). On Helmholtz–Hodge decomposition of inertia on a discrete local frame of reference. *Physics of Fluids*, 32, 083604.

Caltagirone, J.-P. (2021). Application of discrete mechanics model to jump conditions in two-phase flows. *Journal of Computational Physics*, 432, 110151.

Caltagirone, J.-P. (2023). Extension of Galilean invariance to uniform motions for a relativistic equation of fluid flows. *Physics of Fluids*, 35(1), 013103.

Caltagirone, J.-P. (2024). An alternative to the Navier–Stokes equation based on the conservation of acceleration. *Journal of Fluid Mechanics*, 978, A21.

de Gennes, P.G. (1985). Wetting: Statics and dynamics. *Reviews of Modern Physics*, 57, 827–863.

Göhl, J., Mark, A., Sasic, S., Edelvik, F. (2018). An immersed boundary based dynamic contact angle framework for handling complex surfaces of mixed wettabilities. *International Journal of Multiphase Flow*, 109, 164–177.

Gouin, H. and Seppecher, P. (2017). Temperature profile in a liquid–vapor interface near the critical point. *Proceedings of the Royal Society of London. Series A*, 473, 20170229.

Guillaument, R., Vincent, S., Caltagirone, J.-P. (2015). An original algorithm for VOF based method to handle wetting effect in multiphase flow simulation. *Mechanics Research Communications*, 63, 26–32.

Guo, Y., Lian, Y., Sussman, M. (2016). Investigation of drop impact on dry and wet surfaces with consideration of surrounding air. *Physics of Fluids*, 28(7), 073303.

Haley, P. and Miksis, M. (1991). The effect of the contact line on droplet spreading. *Journal of Fluid Mechanics*, 223, 57–81.

Johnson, R.E., Dettre, R.H., Brandreth, D.A. (1977). Dynamic contact angles and contact angle hysteresis. *Journal of Colloid and Interface Science*, 62(2), 205–212.

Kosmann-Schwarzbach, Y. (2011). *Noether Theorems: Invariance and Conservations Laws*. Springer-Verlag, New York.

Legendre, D. and Maglio, M. (2015). Comparison between numerical models for the simulation of moving contact lines. *Computers & Fluids*, 113, 2–13.

Liénard, A. (1898). Champ électrique et magnétique produit par une charge électrique concentrée en un point et animée d'un mouvement quelconque. *L'éclairage électrique*, XVI(27), 5–112.

Maxwell, J. (1865). A dynamical theory of the electromagnetic field. *Philosophical Transactions of the Royal Society of London*, 155, 459–512.

Popinet, S. (2018). Numerical models of surface tension. *Annual Review of Fluid Mechanics*, 50(1), 49–75.

Prosperetti, A. and Tryggvason, G. (2007). *Computational Methods for Multiphase Flows*. Cambridge University Press, Cambridge.

Prud'homme, R. (2012). *Flows and Chemical Reactions*. ISTE Ltd, London and John Wiley & Sons, New York.

Prud'homme, R. (2021). Quadri interfaces. *Thermodynamique des interfaces et mécanique des fluides*, 5(1). doi: 10.21494/ISTE.OP.2021.0665.

Zhang, Z. and Xu, X. (2022). Effective boundary conditions for dynamic contact angle hysteresis on chemically inhomogeneous surfaces. *Journal of Fluid Mechanics*, 935, A34. doi: 10.1017/jfm.2022.33.

9

Numerical Simulation of an Average Flow in an Inter-Blade Channel of a Horizontal Axis Wind Mill

This chapter states a numerical method of finished differences, making it possible to calculate the variables describing an average flow, nonviscous and non-heavy, through a mobile wheel of a windmill with a horizontal axis, to deduce its performances from them, by using a grid with irregular steps. Centered-space discretization diagrams inside the calculation domain and decentered toward the interior for the nodes placed on the borders are used. The temporal discretization uses an explicit diagram with two steps of time, with about two precisions. Calculations are made in two stages: a prediction stage and a correction stage. They preceded calculations of the three-dimensional flow initial state, realized thereafter in its bypass sections. Boundary conditions are imposed on the borders of the calculation domain. Calculation convergence is ensured by numerical viscosity implicitly introduced by the time discretization's diagrams. The internal stability is ensured by a constraint on the step of temporal discretization Δt in accordance with CFL conditions that are calculated while realizing on the flow bypass sections, characteristic quantities initially imposed in the calculation domain volume.

9.1. Introduction

The exploitation of wind energy consists of transforming the kinetic energy of the wind passing through the sensor of a wind turbine [RAJ 23; SMI 22] into mechanical energy recovered at the end of a rotor shaft and then into electrical energy via a generator [TOM 22]. We must therefore start with a study of the wind flow in a site prior to the installation of a wind turbine [BOU 16]. This flow of fluid

---

Chapter written by Delphin TOMBORAVO, Roger VONY, Francis RAVELOSON and Tsialefitry ALY SAANDY.

For a color version of all of the figures in this chapter, see www.iste.co.uk/prudhomme/fluid3.zip.

through a turbomachinery impeller is three-dimensional and unsteady [RAV 19; PEN 11]. It is described by the Euler or Navier–Stokes equation [HAM 10; ELM 19] through systems of nonlinear partial differential equations [SMI 20]. Given the complexity of the analytical solution of these equations, numerical simulation is used, which gives strong results. The magnitude of the computational time for their implementation leads to the search for simpler models for the prediction of turbomachinery performance [SAL 16; CHA 13; DAG 13]. One of these simplified models is the one-dimensional (1D) model.

The present work is oriented towards the calculations of the performance of a wind turbine by considering a 1D average flow equivalent to the 3D flow in the inter-blade channel. For this purpose, the characteristic quantities of the flow are averaged in a passage section perpendicular to its mean direction. To achieve their objectives, the authors have divided their chapter into eight sections. Indeed, the mathematical modeling of fluid motion followed by the definition of the mean flow and the writing of the equations of motion of an average flow in an inter-blade channel opens the theoretical part of the chapter. This part is completed by the calculation of the drive torque of the wheel of a wind turbine, the establishment of the expression of different powers and the calculation of the efficiency. A conclusion reflecting on this perspective ends the chapter.

9.2. Mathematical fluid motion modeling

In an absolute frame (R), the fluid is characterized by its velocity $\vec{V}$, its pressure p, its absolute temperature T and its density ρ.

9.2.1. *Simplifying assumptions*

To simplify the writing of the formulas in this chapter and to differentiate the writing of the mean tensor of a function, the following notations are used:

– $\underline{\underline{\boldsymbol{T}}}$: tensor;

– $\overline{\overline{F}}$: double average of the function;

– $\overline{F}$: simple average of this function.

The fluid is considered to be:

– Newtonian: $\underline{\underline{\boldsymbol{T}}} = p\underline{\underline{\boldsymbol{I}}} + 2\mu\underline{\underline{\boldsymbol{D}}}$ such as $\underline{\underline{\boldsymbol{D}}}$: tensor strain rate;

– $\underline{\underline{\boldsymbol{T}}}$: stress tensor;

– non-weighing: negligible weight;

– nonviscous $\mu = 0$;

– an ideal gas $p = \rho R_g T$;

– thermally conductivity-free: $\vec{q} = \vec{0}$.

The motion of such a fluid is described by Euler's equations established by balances or conservation of mass [9.1], momentum [9.2] and energy [9.3] as follows:

$$\frac{\partial \rho}{dt} + \rho \vec{\nabla} \cdot \vec{V} + \vec{V} \vec{\nabla} \rho = 0 \quad [9.1]$$

$$\frac{\partial \vec{V}}{dt} + \vec{\nabla} \left(\frac{\vec{V}^2}{2} \right) - \vec{V} \wedge \left(\vec{\nabla} \wedge \vec{V} \right) = -\frac{1}{\rho} \vec{\nabla} p \quad [9.2]$$

$$\frac{\partial T}{\partial t} + \frac{a^2}{C_p} \vec{\nabla} \vec{V} + \vec{V} \vec{\nabla} T = 0 \quad [9.3]$$

9.2.2. *Writing equations of motion in a moving coordinate system (E)*

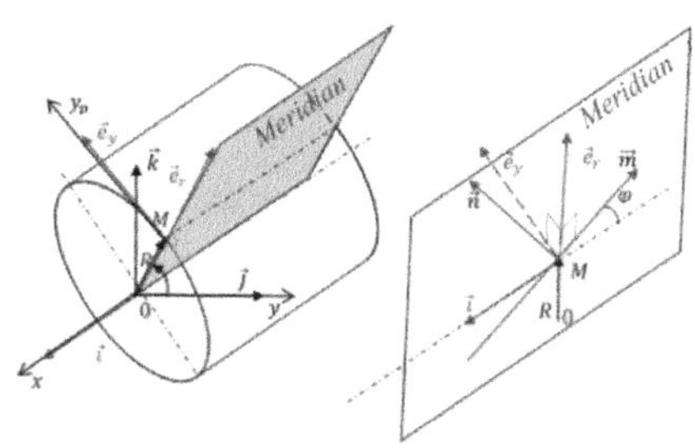

Figure 9.1. *Presentations of (R) and (E) and the angles between their unit vectors*

Figure 9.1 shows the absolute coordinate system $R\left(M, \vec{i}, \vec{j}, \vec{k}\right)$ and the moving marker $E\left(M, \vec{m}, \vec{n}, \vec{e}_y\right)$ with their unit vectors. The unit vector $\vec{i}$ is also the unit vector of the axis of rotation of E with respect to R. In the coordinate system (E), the relative velocity of the flow has the following components: W_m, W_y and W_n.

By designating by ρ', the natural logarithm of the density of the fluid and by $\vec{\omega} \wedge \vec{r}$ the drive speed of (E) with respect to (R) and then setting the rotation vector and $\vec{r} = \overline{OM}$ the position vector, the replacement of $\vec{V}$ by $\vec{W} + \vec{\omega} \wedge \vec{r}$ in equations [9.1]–[9.3] gives the following equations:

$$\frac{\partial_R \rho'}{dt} + \vec{\nabla} \cdot \vec{W} + \vec{W} \vec{\nabla} \rho' = 0 \qquad [9.4]$$

$$\frac{\partial_R \vec{W}}{dt} + \vec{\nabla}\left(\frac{\vec{W}^2}{2}\right) - \vec{W} \wedge \vec{\nabla} \wedge \vec{W} + 2\vec{\omega} \wedge \vec{W} + \vec{\omega} \wedge \vec{\omega} \wedge \vec{r} = -\frac{1}{\rho}\vec{\nabla} p \qquad [9.5]$$

$$\frac{\partial_R T}{\partial t} + \vec{W}\vec{\nabla} T + \frac{a^2}{C_p}\vec{\nabla} \cdot \vec{W} = 0 \qquad [9.6]$$

with $\frac{\partial_R}{dt} = \frac{\partial}{dt} + (\vec{\omega} \wedge \vec{r})\vec{\nabla}$ being the time derivative in the coordinate system (E): equation [9.4] is written as:

$$\begin{aligned} &\frac{\partial_R \rho'}{\partial t} + W_m \frac{\partial \rho'}{\partial m} + W_y \frac{\partial \rho'}{\partial Y} + W_n \frac{\partial \rho'}{\partial n} + \frac{\partial W_m}{\partial m} + \frac{\partial W_y}{\partial y} + \frac{\partial W_n}{\partial n} \\ &+ \left(\frac{\sin\varphi}{R} + \frac{\partial \varphi}{\partial n}\right) W_m + \left(\frac{\cos\varphi}{R} + \frac{\partial \varphi}{\partial m}\right) W_n = 0 \end{aligned} \qquad [9.7]$$

The projection of equation [9.5] on the three axes of the moving frame (E) gives the following equations:

$$\begin{aligned} &\frac{\partial_R W_m}{\partial t} + W_m \frac{\partial W_m}{\partial m} + W_y \frac{\partial W_m}{\partial y} + W_n \frac{\partial W_m}{\partial n} + R_g \frac{\partial T}{\partial m} \\ &+ \frac{a^2}{\gamma}\frac{\partial \rho'}{\partial m} - W_m W_n \frac{\partial \varphi}{\partial m} - W_n^2 \frac{\partial \varphi}{\partial n} - \left(W_y + \omega R\right)^2 \frac{\sin\varphi}{R} = 0 \end{aligned} \qquad [9.8]$$

$$\begin{aligned} &\frac{\partial_R W_y}{\partial t} + W_m \frac{\partial W_y}{\partial m} + W_y \frac{\partial W_y}{\partial y} + W_n \frac{\partial W_y}{\partial n} + R_g \frac{\partial T}{\partial y} \\ &+ \frac{a^2}{\gamma}\frac{\partial \rho'}{\partial y} + \left(W_y + 2\omega R\right)\left(W_m \frac{\sin\varphi}{R} + W_n \frac{\cos\varphi}{R}\right) = 0 \end{aligned} \qquad [9.9]$$

$$\frac{\partial_R W_n}{\partial t}+W_m\frac{\partial W_n}{\partial m}+W_y\frac{\partial W_n}{\partial y}+W_n\frac{\partial W_n}{\partial n}+R_g\frac{\partial T}{\partial n}+\frac{a^2}{\gamma}\frac{\partial \rho'}{\partial n}$$
$$+W_mW_n\frac{\partial \varphi}{\partial n}+W_m^2\frac{\partial \varphi}{\partial m}\left(W_y+\omega R\right)^2\frac{\cos\varphi}{R}=0 \quad [9.10]$$

After the above projection, equation [9.6] becomes as follows:

$$\frac{\partial_R T}{\partial t}+W_m\frac{\partial T}{\partial m}+W_y\frac{\partial T}{\partial y}+W_n\frac{\partial T}{\partial n}+(\gamma-1)$$
$$T\left[\begin{matrix}\dfrac{\partial W_m}{\partial m}+\dfrac{\partial W_y}{\partial y}+\dfrac{\partial W_n}{\partial n}+\left(\dfrac{\sin\varphi}{R}+\dfrac{\partial\varphi}{\partial n}\right)W_m\\ +\left(\dfrac{\cos\varphi}{R}-\dfrac{\partial\varphi}{\partial m}\right)W_n\end{matrix}\right]=0 \quad [9.11]$$

9.3. Definition of average flow

An average flow through an inter-blade channel is a flow whose equations are obtained by using equations [9.7]–[9.11] on a section of passage A perpendicular to the mean direction of this flow is shown in Figure 9.3.

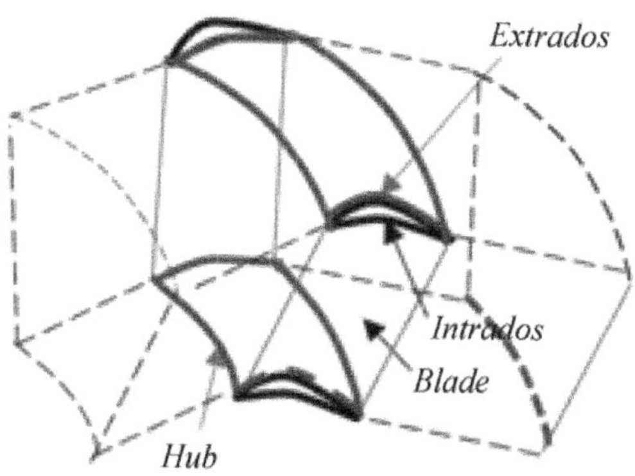

Figure 9.2. *Inter-blade channel*

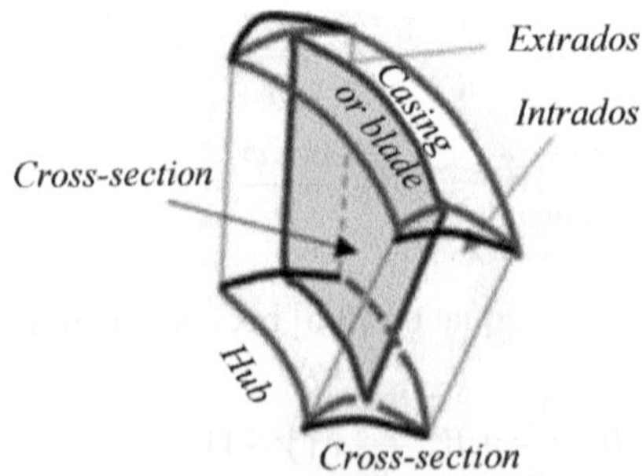

Figure 9.3. *Section of flow passage*

9.3.1. *Definitions of the averages used*

The curvilinear coordinates of the moving frame are m, y *and* n following the direction of $\vec{m}, \vec{e}_y$ and $\vec{n}$. $F(m,y,n)$ is a function of variables m, y and n. Table 9.1 gives the various expressions of the mean of $F(m, y, n)$.

Means F(*m, y, n*)	**Notations and expressions**
Relative to y where (y varying from y_I to y_E)	$\overline{F}^y = \frac{1}{\Delta y}\int_{y_E}^{y_I} F(m,y,n)dy$
Compared to n where n varying from N_S to N_H	$\overline{F}^n = \frac{1}{\Delta n}\int_{N_I}^{N_H} F(m,y,n)dn$
With respect to the section of the passage A perpendicular to the direction m $A = \int_{N_S}^{N_H}\int_{y_E}^{y_I} dydn$	$\overline{\overline{F}} = \frac{1}{A}\int_{N_S}^{N_H}\int_{y_E}^{y_I} F(m,y,n)dydn$
with $\Delta y = y_I - y_E$ and $\Delta n = N_H - N_S$	

Table 9.1. *Expressions of the means of a function* $F(m,y,n)$

In the case of an inter-blade channel, the indices E, I, H and S are indices of quantities taken, respectively, from the upper surface, lower surface, housing and hub.

With $y_I = r_I\theta_I$ and $y_E = r_E\theta_E$ such as r_I *and* r_E radii of the dots on the lower and upper surfaces.

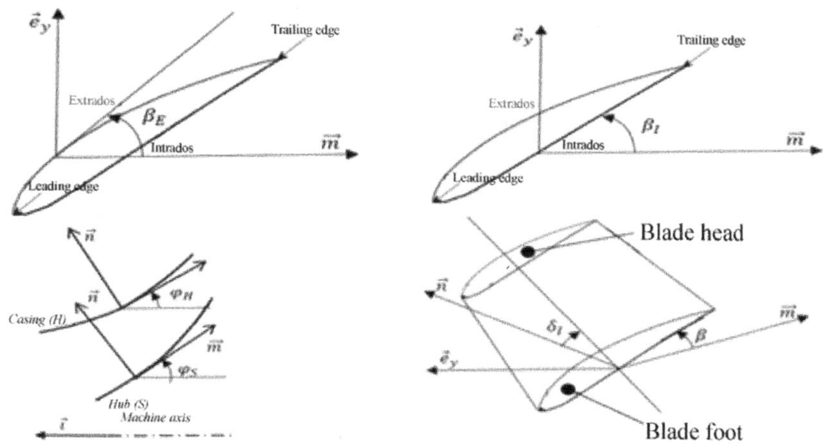

Figure 9.4. *Presentation of the characteristic angles of the channel*

$$\frac{\partial y_E}{\partial m} = \tan\beta_E \qquad \frac{\partial y_I}{\partial m} = \tan\beta_I \qquad \frac{\partial N_S}{\partial m} = \tan\varphi_S \qquad \frac{\partial x}{\partial m} = \cos\varphi$$

$$\frac{\partial y_E}{\partial n} = \tan\delta_E \qquad \frac{\partial y_I}{\partial n} = \tan\delta_I \qquad \frac{\partial N_H}{\partial m} = \tan\varphi_H \qquad \frac{\partial r}{\partial m} = \sin\varphi$$

Means of derivatives with respect to A	Means of derivatives with respect to y	Means of derivatives with respect to n
$\overline{\overline{\frac{\partial F}{\partial m}}} = \frac{\partial \overline{\overline{F}}}{\partial m} + \frac{1}{A}\frac{\partial A}{\partial m}\overline{\overline{F}} + \frac{\Delta n}{A}\left(\tan\beta_E \overline{F_E}^n - \tan\beta_I \overline{F_I}^n\right) + \frac{\Delta y}{A}\left(\tan\varphi_S \overline{F_S}^y - \tan\varphi_H \overline{F_H}^y\right)$	$\overline{\frac{\partial F}{\partial m}}^y = \frac{\partial \overline{F}^y}{\partial m} + \frac{1}{\Delta y}\frac{\partial \Delta y}{\partial m}\overline{F}^y + \frac{1}{\Delta y}\left(\tan\delta_E F_E - \tan\delta_I F_I\right)$	$\overline{\frac{\partial F}{\partial m}}^n = \frac{\partial \overline{F}^n}{\partial m} + \frac{1}{\Delta n}\frac{\partial \Delta n}{\partial m}\overline{F}^n + \frac{1}{\Delta n}\left(\tan\varphi_S F_S - \tan\varphi_H F_H\right)$
$\overline{\overline{\frac{\partial F}{\partial n}}} = \frac{\Delta y}{A}\left(\overline{F_H}^y - \overline{F_S}^y\right)$	$\overline{\frac{\partial F}{\partial n}}^y = \frac{\partial \overline{F}^y}{\partial n} + \frac{1}{\Delta y}\frac{\partial \Delta y}{\partial n}\overline{F}^y + \frac{1}{\Delta y}\left(\tan\delta_E F_E - \tan\delta_I F_I\right)$	$\overline{\frac{\partial F}{\partial n}}^n = \frac{1}{\Delta n}\left(F_H - F_S\right)$
$\overline{\overline{\frac{\partial F}{\partial y}}} = \frac{\Delta y}{A}\left(\overline{F_I}^n - \overline{F_E}^n\right)$	$\overline{\frac{\partial F}{\partial y}}^y = \frac{1}{\Delta y}\left(F_I - F_E\right)$	$\overline{\frac{\partial F}{\partial y}}^n = \frac{\partial \overline{F}^n}{\partial y} + \frac{1}{\Delta n}\frac{\partial \Delta n}{\partial y}\overline{F}^n + \frac{1}{\Delta n}\left(\tan\varphi_S F_S - \tan\varphi_H F_H\right)$

Table 9.2. *Expressions of the means of spatial derivatives of a function F*

In accordance with the definitions of means, angles and characteristic quantities of the surfaces delimiting the volume in which the flow evolves above, the expressions of the means of spatial derivatives of the function $F\ (m, y, n)$, for the double average $\overline{\overline{F}}$ and for the simple average $\overline{F}$, are summarized in Table 9.2.

9.4. Writing the equations of motion of a mean flow in an inter-blade channel

It is assumed that the values of the characteristic quantities of the flow remain close to those of the mean flow. Using this in Table 9.2, the averages over the cross section A of equations [9.7]–[9.11] are given by the following equations.

These equations contain the average quantities on the upper surface, lower surface, crankcase and hub. The equations that give their temporal variations are described in the subsequent sections.

$$
\begin{aligned}
&\overline{\overline{\frac{\partial_R \rho'}{\partial t}}} + \overline{\overline{W_m}}\,\overline{\overline{\frac{\partial \rho'}{\partial m}}} + \overline{\overline{W_m}}\,\overline{\overline{\frac{\partial m}{\partial m}}} + \frac{1}{A}\frac{\partial A}{\partial m}\overline{\overline{\rho' W_m}} \\
&+\left[\begin{array}{l}\dfrac{1}{A}\dfrac{\partial A}{\partial m} + \dfrac{\Delta n}{A}\left(\overline{\rho' E}^{n}\,\overline{\tan\beta_E}^{n} - \overline{\rho' I}^{n}\,\overline{\tan\beta_I}^{n}\right) + \dfrac{\Delta y}{A} \\ \left(\dfrac{\partial N_S}{\partial m}\overline{\rho' S}^{y} - \dfrac{\partial N_H}{\partial m}\overline{\rho' H}^{y}\right) + \overline{\overline{\dfrac{\partial \theta}{\partial Y}}}\,\overline{\overline{\sin\varphi}} + \overline{\overline{\dfrac{\partial \varphi}{\partial n}}}\end{array}\right] W_m \\
&+\left[\frac{\Delta y}{A}\left(\overline{\rho' H}^{y} - \overline{\rho' S}^{y} + \overline{\overline{\frac{\partial \theta}{\partial Y}}}\,\overline{\overline{\cos\varphi}} - \overline{\overline{\frac{\partial \varphi}{\partial m}}}\right)\right] W_n + \frac{\Delta n}{A}\left(\overline{\rho' H}^{y} - \overline{\rho' S}^{y}\right) W_y \\
&+\frac{\Delta n}{A}\left(\overline{W_{nE}}^{n}\,\overline{\tan\beta_E}^{n} - \overline{W_{nI}}^{n}\,\overline{\tan\beta_I}^{n} + \overline{W_{yI}}^{n} - \overline{W_{yE}}^{n}\right) \\
&+\frac{\Delta y}{A}\left(\frac{\partial N_S}{\partial m}\overline{W_{mS}}^{y} - \frac{\partial N_H}{\partial m}\overline{W_{mH}}^{y} + \overline{W_{nH}}^{y} - \overline{W_{nS}}^{y}\right) = 0
\end{aligned}
\tag{9.12}
$$

$$\begin{aligned}
&\frac{\overline{\overline{\partial_R W_m}}}{\partial t} + \overline{\overline{W_m}}\frac{\partial \overline{\overline{W_m}}}{\partial m} + R_g \overline{\overline{T}}\frac{\partial \overline{\overline{\rho'}}}{\partial m} + R_g \frac{\partial \overline{\overline{T}}}{\partial m} + R_g \frac{1}{A}\frac{\partial A}{\partial m}\overline{\overline{\rho' T}} + \frac{1}{A}\frac{\partial A}{\partial m}\overline{\overline{W_m^2}} - \overline{\overline{W_n^2}}\frac{\overline{\overline{\partial \varphi}}}{\partial n} \\
&\quad - \overline{\overline{W_y^2}}\left(\frac{\overline{\overline{\partial \theta}}}{\partial Y}\overline{\overline{R}} - 1\right)\frac{\overline{\overline{\sin\varphi}}}{R} - \left(\overline{\overline{W_y}} + \overline{\overline{U}}\right)^2 \frac{\overline{\overline{\sin\varphi}}}{R} - \frac{\partial \varphi}{\partial m}\overline{\overline{W_m W_n}} \\
&\quad + \left[\begin{array}{l} \dfrac{\Delta n}{A}\left(\overline{W_{mE}}^n\,\overline{\tan\beta_E}^n - \overline{W_{mI}}^n\,\overline{\tan\beta_I}^n\right) \\ + \dfrac{\Delta y}{A}\left(\dfrac{\partial N_S}{\partial m}\overline{W_{mS}}^y - \dfrac{\partial N_H}{\partial m}\overline{W_{mH}}^y\right) \end{array}\right]\overline{\overline{W_m}} \\
&\quad + \left[\frac{\Delta n}{A}\left(\overline{W_{mI}}^n - \overline{W_{mE}}^n\right)\right]\overline{\overline{W_y}} + \left[\frac{\Delta y}{A}\left(\overline{W_{mH}}^n - \overline{W_{mS}}^n\right)\right]\overline{\overline{W_n}} \\
&\quad + R_g\left[\begin{array}{l} \dfrac{1}{A}\dfrac{\partial A}{\partial m} + \dfrac{\Delta n}{A}\left(\overline{\rho'_E}^n\,\overline{\tan\beta_E}^n - \overline{\rho'_I}^n\,\overline{\tan\beta_I}^n\right) \\ + \dfrac{\Delta y}{A}\left(\dfrac{\partial N_S}{\partial m}\overline{\rho'_S}^y - \dfrac{\partial N_H}{\partial m}\overline{\rho'_H}^y\right) \end{array}\right]\overline{\overline{T}} \\
&\quad + R_g\left[\begin{array}{l} \dfrac{\Delta n}{A}\left(\overline{T_E}^n\,\overline{\tan\beta_E}^n - \overline{T_I}^n\,\overline{\tan\beta_I}^n\right) \\ + \dfrac{\Delta y}{A}\left(\dfrac{\partial N_S}{\partial m}\overline{T_S}^y - \dfrac{\partial N_H}{\partial m}\overline{T_H}^y\right) \end{array}\right] = 0
\end{aligned} \qquad [9.13]$$

$$\begin{aligned}
&\frac{\overline{\overline{\partial_R W_y}}}{\partial t} + \overline{\overline{W_m}}\frac{\partial \overline{\overline{W_y}}}{\partial m} + \frac{1}{A}\frac{\partial A}{\partial m}\overline{\overline{W_y W_m}} + \frac{\overline{\overline{\sin\varphi}}}{R}\left(\overline{\overline{W_y}}\frac{\overline{\overline{\partial\theta}}}{\partial Y}\overline{\overline{R}} + 2\overline{\overline{U}}\right)\overline{\overline{W_m}} \\
&\quad + \frac{\overline{\overline{\cos\varphi}}}{R}\left(\overline{\overline{W_y}}\frac{\overline{\overline{\partial\theta}}}{\partial Y}\overline{\overline{R}} + 2\overline{\overline{U}}\right)\overline{\overline{W_n}} \\
&\quad + \left[\begin{array}{l} -\dfrac{\Delta n}{A}\left(\overline{W_{yI}}^n\,\overline{\tan\beta_I}^n - \overline{W_{yE}}^n\,\overline{\tan\beta_E}^n\right) \\ + \dfrac{\Delta y}{A}\left(\dfrac{\partial N_S}{\partial m}\overline{W_{yS}}^y - \dfrac{\partial N_H}{\partial m}\overline{W_{yH}}^y\right) \end{array}\right]\overline{\overline{W_m}} \\
&\quad + \frac{\Delta n}{A}\left(\overline{W_{yI}}^n - \overline{W_{yE}}^n\right)\overline{\overline{W_y}} + \frac{\Delta y}{A}\left(\overline{W_{yH}}^y - \overline{W_{yS}}^y\right)\overline{\overline{W_n}} \\
&\quad + R_g\frac{\Delta n}{A}\left(\overline{\rho'_I}^n - \overline{\rho'_E}^n\right)\overline{\overline{T}} + R_g\frac{\Delta n}{A}\left(\overline{T_I}^n - \overline{T_E}^n\right) = 0
\end{aligned} \qquad [9.14]$$

$$\begin{aligned}
&\frac{\overline{\overline{\partial_R W_n}}}{\partial t}+\overline{\overline{W_m}}\frac{\partial\overline{\overline{W_n}}}{\partial m}+\overline{\overline{W_m^2}}\frac{\partial\varphi}{\partial m}+\left(\frac{1}{A}\frac{\partial A}{\partial m}+\frac{\partial\varphi}{\partial n}\right)\overline{\overline{W_n W_m}}\\
&-\left(\overline{\overline{W_y}}+\overline{\overline{U}}\right)^2\overline{\overline{\frac{\cos\varphi}{R}}}-\overline{\overline{W_y^2}}\left(\frac{\partial\overline{\overline{\theta}}}{\partial Y}\overline{\overline{R}}-1\right)\overline{\overline{\frac{\cos\varphi}{R}}}\\
&\quad+\left[\frac{\Delta n}{A}\left(\overline{W_{nE}}^{n}\,\overline{\tan\beta_E}^{n}-\overline{W_{nI}}^{n}\,\overline{\tan\beta_I}^{n}\right)+\frac{\Delta y}{A}\left(\frac{\partial N_S}{\partial m}\overline{W_{nS}}^{y}-\frac{\partial N_H}{\partial m}\overline{W_{nH}}^{y}\right)\right]\\
&\quad\overline{\overline{W_m}}+\frac{\Delta n}{A}\left(\overline{W_{nI}}^{n}-\overline{W_{nE}}^{n}\right)\overline{\overline{W_y}}\\
&\quad+\frac{\Delta y}{A}\left(\overline{W_{nH}}^{y}-\overline{W_{nS}}^{y}\right)\overline{\overline{W_n}}+R_g\frac{\Delta y}{A}\left(\overline{\rho'_H}^{y}-\overline{\rho'_S}^{y}\right)\overline{\overline{T}}\\
&\quad+R_g\frac{\Delta y}{A}\left(\overline{T_H}^{y}-\overline{T_S}^{y}\right)=0
\end{aligned}\qquad [9.15]$$

$$\begin{aligned}
&\frac{\partial_R\overline{\overline{T}}}{\partial t}+\overline{\overline{W_m}}\frac{\partial\overline{\overline{T}}}{\partial m}+(\gamma-1)\overline{\overline{T}}\frac{\partial\overline{\overline{W_m}}}{\partial m}+(\gamma-1)\left(\frac{\partial\overline{\overline{\theta}}}{\partial Y}\cos\varphi-\frac{\partial\overline{\overline{\varphi}}}{\partial m}\right)\overline{\overline{TW_n}}\\
&\quad+\left[\gamma\frac{1}{A}\frac{\partial A}{\partial m}+(\gamma-1)\left(\frac{\partial\overline{\overline{\theta}}}{\partial Y}\sin\varphi-\frac{\partial\overline{\overline{\varphi}}}{\partial n}\right)\right]\overline{\overline{TW_m}}+\frac{\Delta n}{A}\left(\overline{T_I}^{n}-\overline{T_E}^{n}\right)\overline{\overline{W_y}}\\
&\quad+\frac{\Delta y}{A}\left(\overline{T_H}^{y}-\overline{T_S}^{y}\right)\overline{\overline{W_n}}\\
&\quad+\left[\frac{\Delta n}{A}\left(\overline{T_E}^{n}\,\overline{\tan\beta_E}^{n}-\overline{T_I}^{n}\,\overline{\tan\beta_I}^{n}\right)+\frac{\Delta y}{A}\left(\frac{\partial N_S}{\partial m}\overline{T_S}^{y}-\frac{\partial N_H}{\partial m}\overline{T_H}^{y}\right)\right]\overline{\overline{W_m}}\\
&\quad+(\gamma-1)\left[\begin{aligned}&\frac{\Delta n}{A}\left(\overline{W_{mE}}^{n}\,\overline{\tan\beta_E}^{n}-\overline{W_{mI}}^{n}\,\overline{\tan\beta_I}^{n}\right)\\&+\frac{\Delta y}{A}\left(\frac{\partial N_S}{\partial m}\overline{W_{mS}}^{y}-\frac{\partial N_H}{\partial m}\overline{W_{mH}}^{y}\right)\\&+\frac{\Delta y}{A}\left(\overline{W_{nH}}^{y}-\overline{W_{nS}}^{y}\right)+\frac{\Delta n}{A}\left(\overline{W_{yI}}^{n}-\overline{W_{yE}}^{n}\right)\end{aligned}\right]\overline{\overline{T}}=0
\end{aligned}\qquad [9.16]$$

9.4.1. *Expressions of the equations of motion of the mean flows on the hub and on the crankcase*

The hub and the crankcase coincide with surfaces corresponding to $n = N_H$ and $n = N_S$.

To have the expressions of the average equations of motion on the housing and on the hub, we must average the equations of motion of [9.7]–[9.11] according to y, which gives:

$$
\begin{aligned}
&\frac{\partial_R \overline{\rho'_W}^y}{\partial t} + \overline{W_{mW}}^y \frac{\partial \overline{\rho'_W}^y}{\partial m} + \overline{W_{nW}}^y \frac{\partial \overline{\rho'_W}^y}{\partial n} + \frac{\partial \overline{W_{mW}}^y}{\partial m} + \frac{\partial \overline{W_{nW}}^y}{\partial n} \\
&+ \frac{1}{\Delta y}\frac{\partial \Delta y}{\partial n} \overline{\rho'_W}^y \overline{W_{nW}}^y + \frac{1}{\Delta y}\frac{\partial \Delta y}{\partial m} \overline{\rho'_W}^y \overline{W_{mW}}^y \\
&+ \frac{1}{\Delta y}\left(\rho'_{WI} - \rho'_{WE}\right)\overline{W_{yW}}^y + \left[\begin{array}{l} \dfrac{1}{\Delta y}\left(\tan\beta_{WE}\rho'_{WE} - \tan\beta_{WI}\rho'_{WI} + \dfrac{\partial \Delta y}{\partial m}\right) \\ + \overline{\left(\dfrac{\partial \theta}{\partial Y}\right)_W}^y \sin\varphi_W + \dfrac{\partial \varphi_W}{\partial n} \end{array}\right] \overline{W_{mW}}^y \qquad [9.17] \\
&+ \left[\frac{1}{\Delta y}\left(\tan\delta_{WE}\rho'_{WE} - \tan\delta_{WI}\rho'_{WI} + \frac{\partial \Delta y}{\partial n}\right) + \overline{\left(\frac{\partial \theta}{\partial Y}\right)_W}^y \cos\varphi_W - \frac{\partial \varphi_W}{\partial m}\right] \\
&\overline{W_{nW}}^y + \frac{1}{\Delta y}\left(W_{yWI} - W_{yWE}\right) \\
&+ \frac{1}{\Delta y}\left(\tan\beta_{WE}W_{mWE} - \tan\beta_{WI}W_{mWI} + \tan\delta_{WE}W_{nWE} - \tan\delta_{WI}W_{nWI}\right) = 0
\end{aligned}
$$

$$
\begin{aligned}
&\frac{\overline{\partial_R W_{mW}}^y}{\partial t} + \overline{W_{mW}}^y \frac{\partial \overline{W_{mW}}^y}{\partial m} + \overline{W_{nW}}^y \frac{\partial \overline{W_{mW}}^y}{\partial n} + R_g \overline{T_W}^y \frac{\partial \overline{\rho'_W}^y}{\partial n} \\
&+R_g \frac{\partial \overline{T_W}^y}{\partial n} + \frac{1}{\Delta y} R_g \frac{\partial \Delta y}{\Delta m} \overline{\rho'_W}^y \overline{T_W}^y \\
&\quad + \frac{1}{\Delta y}\frac{\partial \Delta y}{\Delta m}\overline{W_{mW}^2}^y - \frac{\partial \overline{\varphi_W}^y}{\partial n}\overline{W_{nW}^2}^y - \frac{\overline{\sin\varphi_W}^y}{R_W}\left(\overline{W_{yW}}^y - \omega R_W\right)^2 \\
&\quad - \frac{\overline{\sin\varphi_W}^y}{R_W}\left[\overline{\left(\frac{\partial \theta}{\partial Y}\right)_W}^y \overline{R_W}^y - 1\right]\overline{W_{yW}^2}^y \\
&\quad + \left(\frac{1}{\Delta y}\frac{\partial \Delta y}{\Delta n} - \frac{\partial \overline{\varphi_W}^y}{\partial m}\right)\overline{W_{mW}}^y \overline{W_{nW}}^y \\
&\quad + \frac{1}{\Delta y}\left(\tan\beta_{WE} W_{mWE} - \tan\beta_{WI} W_{mWI}\right)\overline{W_{mW}}^y \\
&\quad + \frac{1}{\Delta y}\left(W_{mWI} - W_{mWE}\right)\overline{W_{yW}}^y + \frac{1}{\Delta y}\left(\tan\delta_E W_{mWE} - \tan\delta_I W_{mWI}\right)\overline{W_{nW}}^y \\
&\quad + \frac{1}{\Delta y}\left(\frac{\partial \Delta y}{\Delta m} + \tan\beta_{WE}\rho'_{WE} - \tan\beta_{WI}\rho'_{WI}\right)R_g \overline{T_W}^y \\
&\quad + \frac{1}{\Delta y} Rg\left(\tan\beta_{WE} T_{WE} - \tan\beta_{WI} T_{WI}\right) = 0
\end{aligned}
\qquad [9.18]
$$

$$
\begin{aligned}
&\frac{\overline{\partial_R W_{yW}}^{y}}{\partial t} + \overline{W_{mW}}^{y}\frac{\overline{\partial W_{yW}}^{y}}{\partial m} + \overline{W_{nW}}^{y}\frac{\overline{\partial W_{yW}}^{y}}{\partial n} + \frac{1}{\Delta y}\frac{\partial \Delta y}{\Delta m}\overline{W_{yW}}^{y}\overline{W_{mW}}^{y} \\
&+\frac{1}{\Delta y}\frac{\partial \Delta y}{\Delta n}\overline{W_{yW}}^{y}\overline{W_{nW}}^{y} \\
&\quad+\left[\overline{W_{yW}}^{y}\overline{\left(\frac{\partial \theta}{\partial Y}\right)_{W}}^{y}\overline{R_W}^{y} - 2\omega R_W\right]\left(\overline{W_{mW}}^{y}\frac{\overline{\sin\varphi_W}^{y}}{R_W} + \overline{W_{nW}}^{y}\frac{\overline{\cos\varphi_W}^{y}}{R_W}\right) \\
&\quad+\frac{1}{\Delta y}\left(W_{yWI} - W_{yWE}\right)\overline{W_{yW}}^{y} \qquad [9.19] \\
&\quad+\frac{1}{\Delta y}\left(\tan\beta_{WE}W_{yWE} - \tan\beta_{WI}W_{yWI}\right)\overline{W_{mW}}^{y} \\
&\quad+\frac{1}{\Delta y}\left(\tan\delta_{E}W_{yWE} - \tan\delta_{I}W_{yWI}\right)\overline{W_{nW}}^{y} \\
&\quad+R_g\frac{1}{\Delta y}\left(\rho'_{WI} - \rho'_{WE}\right)\overline{T_W}^{y} + Rg\frac{1}{\Delta y}\left(T_{WI} - T_{WE}\right) = 0
\end{aligned}
$$

$$
\begin{aligned}
&\frac{\overline{\partial_R W_{nW}}^y}{\partial t} + \overline{W_{mW}}^y \frac{\overline{\partial W_{nW}}^y}{\partial m} + \overline{W_{nW}}^y \frac{\overline{\partial W_{nW}}^y}{\partial n} + R_g \overline{T_W}^y \frac{\overline{\partial \rho'_W}^y}{\partial n} + R_g \frac{\overline{\partial T_W}^y}{\partial n} \\
&+ \frac{1}{\Delta y} R_g \frac{\partial \Delta y}{\Delta n} \overline{\rho'_W}^y \overline{T_W}^y \\
&\qquad + \frac{\overline{\partial \varphi_W}^y}{\partial m} \overline{W_{mW}^2}^y + \frac{1}{\Delta y} \frac{\partial \Delta y}{\partial n} \overline{W_{nW}^2}^y - \left(\overline{W_{yW}}^y - \omega R \right)^2 \frac{\overline{\cos \varphi_W}^y}{R_W} - \overline{W_{yW}^2}^y \\
&\left[\overline{\left(\frac{\partial \theta}{\partial Y} \right)_W}^y \overline{R_W}^y - 1 \right] \frac{\overline{\cos \varphi_W}^y}{R_W} \\
&\qquad + \left(\frac{1}{\Delta y} \frac{\partial \Delta y}{\Delta m} + \frac{\overline{\partial \varphi_W}^y}{\partial n} \right) \overline{W_{nW}}^y \overline{W_{mW}}^y \\
&+ \frac{1}{\Delta y} \left(\tan \beta_{WE} W_{nWE} - \tan \beta_{WI} W_{nWI} \right) \overline{W_{mW}}^y \\
&\qquad + \frac{1}{\Delta y} \left(W_{nWI} - W_{nWE} \right) \overline{W_{yW}}^y + \frac{1}{\Delta y} \left(\tan \delta_E W_{nWE} - \tan \delta_I W_{nWI} \right) \overline{W_{nW}}^y \\
&\qquad + \frac{1}{\Delta y} R_g \left(\frac{\partial \Delta y}{\Delta n} + \tan \delta_{WE} W_{nWE} - \tan \delta_{WI} W_{nWI} \right) \overline{T_W}^y \\
&\qquad + \frac{1}{\Delta y} Rg \left(\tan \delta_{WE} T_{WE} - \tan \delta_{WI} T_{WI} \right) = 0
\end{aligned}
\qquad [9.20]
$$

$$
\begin{aligned}
&\frac{\partial_R \overline{T_W}^y}{\partial t} + \overline{W_{mW}}^y \frac{\partial \overline{T_W}^y}{\partial m} + \overline{W_{nW}}^y \frac{\partial \overline{T_W}^y}{\partial n} + (\gamma-1)\overline{T_W}^y \frac{\partial \overline{W_{mW}}^y}{\partial m} \\
&+(\gamma-1)\overline{T_W}^y \frac{\partial \overline{W_{nW}}^y}{\partial n} \\
&\quad + \left\{ \gamma \frac{1}{\Delta y}\frac{\partial \Delta y}{\partial m}\overline{W_{mW}}^y + (\gamma-1)\left[\overline{\left(\frac{\partial \theta}{\partial Y}\right)_W}^y \sin\varphi_W + \frac{\partial \overline{\varphi_W}^y}{\partial n} \right] \right\} \overline{T_W}^y \\
&\quad + \left\{ \gamma \frac{1}{\Delta y}\frac{\partial \Delta y}{\partial n} + (\gamma-1)\left[\overline{\left(\frac{\partial \theta}{\partial Y}\right)_W}^y \cos\varphi_W + \frac{\partial \overline{\varphi_W}^y}{\partial m} \right] \right\} \overline{W_{nW}}^y \overline{T_W}^y \qquad [9.21] \\
&\quad + \frac{1}{\Delta y}\left(\tan\delta_E T_{WE} - \tan\delta_I T_{WI}\right)\overline{W_{nW}}^y \\
&\quad + \frac{1}{\Delta y}(\gamma-1)\left(\tan\beta_{WE} T_{WE} - \tan\beta_{WI} T_{WI}\right)\overline{W_{mW}}^y + \frac{1}{\Delta y}\left(T_{WI} - T_{WE}\right)\overline{W_{yW}}^y \\
&\quad + \frac{1}{\Delta y}(\gamma-1)\begin{pmatrix} \tan\beta_{WE} W_{mWE} - \tan\beta_{WI} W_{mWI} + \tan\delta_{WE} W_{nWE} \\ -\tan\delta_{WI} W_{nWI} + W_{yWI} - W_{yWE} \end{pmatrix} \overline{T_W}^y = 0
\end{aligned}
$$

The index W is replaced by H for the case of the average flow on the crankcase and replaced by S for the case of the average flow on the hub. E and I are the indices of the quantities taken from the upper and lower surfaces.

9.4.2. *Expressions of the equations of motion of the mean flows on the upper surface and on the lower surface*

To obtain the expressions of the average equations of motion on the upper surface and on the lower surface, the equations of motion of [9.7]–[9.11] must be averaged according to n. This gives:

$$
\begin{aligned}
&\frac{\partial_R \overline{\rho'_W}^n}{\partial t} + \overline{W_{mW}}^n \frac{\partial \overline{\rho'_W}^n}{\partial m} + \overline{W_{yW}}^n \frac{\partial \overline{\rho'_W}^n}{\partial y} + \frac{\partial \overline{W_{mW}}^n}{\partial m} + \frac{\partial \overline{W_{yW}}^n}{\partial y} \\
&+ \frac{1}{\Delta n}\frac{\partial \Delta n}{\partial m}\overline{\rho'_W}^n\,\overline{W_{mW}}^n + \frac{1}{\Delta n}\frac{\partial \Delta n}{\partial y}\overline{\rho'_W}^n\,\overline{W_{yW}}^n \\
&\quad + \left[\begin{array}{l} \dfrac{1}{\Delta n}\left(\dfrac{\partial N_S}{\partial m}\rho'_{WS} - \dfrac{\partial N_H}{\partial m}\rho'_{WH} + \dfrac{\partial \Delta n}{\partial m}\right) \\ + \overline{\left(\dfrac{\partial \theta}{\partial Y}\right)_W}^n \sin\varphi_W + \dfrac{\partial \varphi_W}{\partial n} \end{array}\right]\overline{W_{mW}}^n \\
&\quad + \frac{1}{\Delta n}\left(\frac{\partial N_S}{\partial m}\rho'_{WS} - \frac{\partial N_H}{\partial m}\rho'_{WH} + \frac{\partial \Delta n}{\partial y}\right)\overline{W_{yW}}^n \\
&+ \left[\frac{1}{\Delta n}\left(\rho'_{WH} - \rho'_{WS}\right) + \overline{\left(\frac{\partial \theta}{\partial Y}\right)_W}^n \cos\varphi_W - \frac{\partial \varphi_W}{\partial m}\right]\overline{W_{nW}}^n \\
&\quad + \frac{1}{\Delta n}\left(\begin{array}{l} W_{nWH} - W_{nWS} + \dfrac{\partial N_S}{\partial m}W_{mWS} - \dfrac{\partial N_H}{\partial m}W_{mWH} \\ + \dfrac{\partial N_S}{\partial y}W_{yWS} - \dfrac{\partial N_H}{\partial y}W_{yWH} \end{array}\right) = 0
\end{aligned}
\qquad [9.22]
$$

$$\begin{aligned}
&\frac{\overline{\partial_R W_{mW}}^n}{\partial t}+\overline{W_{mW}}^n\frac{\overline{\partial W_{mW}}^n}{\partial m}+\overline{W_{yW}}^n\frac{\overline{\partial W_{mW}}^n}{\partial y}+R_g\frac{\overline{\partial T_W}^n}{\partial m}\\
&+R_g\overline{T_W}^n\frac{\overline{\partial \rho'_W}^n}{\partial m}+\frac{1}{\Delta n}R_g\frac{\partial\Delta n}{\partial m}\overline{T_W}^n\overline{\rho'_W}^n\\
&\quad+\frac{1}{\Delta n}\frac{\partial\Delta n}{\partial m}\overline{W_{mW}^2}^n-\overline{W_{nW}^2}^n\frac{\overline{\partial\varphi_W}^n}{\partial n}-\frac{\overline{\sin\varphi_W}^n}{R_W}\left(\overline{W_{yW}}^n+\omega R_W\right)^2\\
&-\overline{W_{yW}^2}^n\left[\overline{\left(\frac{\partial\theta}{\partial Y}\right)_W}^n R_W-1\right]\frac{\overline{\sin\varphi_W}^n}{R_W}\\
&\quad+\frac{1}{\Delta n}\frac{\partial\Delta n}{\partial m}\overline{W_{mW}}^n\overline{W_{yW}}^n-\frac{\overline{\partial\varphi_W}^n}{\partial m}\overline{W_{mW}}^n\overline{W_{nW}}^n\\
&+\frac{1}{\Delta n}\left(\frac{\partial N_S}{\partial m}\overline{W_{mWS}}^n-\frac{\partial N_H}{\partial m}\overline{W_{mWH}}^n\right)\overline{W_{mW}}^n\\
&\quad+\frac{1}{\Delta n}\left(\frac{\partial N_S}{\partial y}\overline{W_{mWS}}^n-\frac{\partial N_H}{\partial y}\overline{W_{mWH}}^n\right)\overline{W_{yW}}^n\\
&+\frac{1}{\Delta n}\left(W_{mWH}-W_{mWS}\right)\overline{W_{nW}}^n\\
&\quad+\frac{1}{\Delta n}R_g\left(\frac{\partial\Delta n}{\partial m}+\frac{\partial N_S}{\partial m}\overline{\rho'_{WS}}^n-\frac{\partial N_H}{\partial m}\overline{\rho'_{WH}}^n\right)\overline{T_W}^n\\
&+\frac{1}{\Delta n}R_g\left(\frac{\partial N_S}{\partial m}\overline{T_{WS}}^n-\frac{\partial N_H}{\partial m}\overline{T_{WH}}^n\right)=0
\end{aligned}\qquad [9.23]$$

$$
\begin{aligned}
&\frac{\overline{\partial_R W_{nW}}^n}{\partial t} + \overline{W_{mW}}^n \frac{\overline{\partial W_{nW}}^n}{\partial m} + \overline{W_{yW}}^n \frac{\overline{\partial W_{nW}}^n}{\partial y} + \frac{\overline{\partial \varphi_W}^n}{\partial m} \overline{W_{mW}^2}^n \\
&- \frac{\overline{\cos\varphi_W}^n}{R_W}\left(\overline{W_{yW}}^n + \omega \overline{R_W}^n\right)^2 \\
&\qquad - \frac{\overline{\cos\varphi_W}^n}{R_W}\left[\overline{\left(\frac{\partial \theta}{\partial Y}\right)_W R_W - 1}^n\right]\overline{W_{yW}^2}^n + \left(\frac{1}{\Delta n}\frac{\partial \Delta n}{\partial m} + \frac{\overline{\partial \varphi_W}^n}{\partial n}\right)\overline{W_{nW}}^n \overline{W_{mW}}^n \\
&+ \frac{1}{\Delta n}\frac{\partial \Delta n}{\partial y}\overline{W_{nW}}^n \overline{W_{yW}}^n \qquad\qquad [9.24] \\
&\qquad + \frac{1}{\Delta n}\left(\frac{\partial N_S}{\partial m}\overline{W_{nWS}}^n - \frac{\partial N_H}{\partial m}\overline{W_{nWH}}^n\right)\overline{W_{mW}}^n \\
&+ \frac{1}{\Delta n}\left(\frac{\partial N_S}{\partial y}\overline{W_{nWS}}^n - \frac{\partial N_H}{\partial y}\overline{W_{nWH}}^n\right)\overline{W_{yW}}^n \\
&\qquad + \frac{1}{\Delta n}\left(W_{nWH} - W_{nWS}\right)\overline{W_{nW}}^n + \frac{1}{\Delta n} R_g\left(\rho'_{WH} - \rho'_{WS}\right)\overline{T_W}^n \\
&+ \frac{1}{\Delta n} R_g\left(T_{WH} - T_{WS}\right) = 0
\end{aligned}
$$

$$
\begin{aligned}
&\frac{\overline{\partial_R W_{yW}}^n}{\partial t} + \overline{W_{mW}}^n \frac{\overline{\partial W_{yW}}^n}{\partial m} + \overline{W_{yW}}^n \frac{\overline{\partial W_{yW}}^n}{\partial y} + R_g \frac{\overline{\partial T_W}^n}{\partial y} \\
&+R_g \overline{T_W}^n \frac{\overline{\partial \rho'_W}^n}{\partial y} + \frac{1}{\Delta n} R_g \frac{\partial \Delta n}{\partial m} \overline{T_W}^n \overline{\rho'_W}^n \\
&\quad + \frac{1}{\Delta n}\frac{\partial \Delta n}{\partial m} \overline{W_{yW}{}^2}^n + \left[\overline{W_{yW}}^n \overline{\left(\frac{\partial \theta}{\partial Y}\right)_W}^n \overline{R_W}^n + 2\omega \overline{R_W}^n \right] \\
&\left(\overline{W_{mW}}^n \frac{\overline{\sin \varphi_W}^n}{R_W} - \overline{W_{nW}}^n \frac{\overline{\cos \varphi_W}^n}{R_W} \right) \\
&\quad + \frac{1}{\Delta n}\frac{\partial \Delta n}{\partial m} \overline{W_{yW}}^n \overline{W_{mW}}^n + \frac{1}{\Delta n}\left(\frac{\partial N_S}{\partial y} \overline{W_{yWS}}^n - \frac{\partial N_H}{\partial y} \overline{W_{yWH}}^n \right) \overline{W_{mW}}^n \qquad [9.25] \\
&+ \frac{1}{\Delta n} (W_{mWH} - W_{mWS}) \overline{W_{nW}}^n \\
&\quad + \frac{1}{\Delta n}\left(\frac{\partial N_S}{\partial y} \overline{W_{yWS}}^n - \frac{\partial N_H}{\partial y} \overline{W_{yWH}}^n \right) \overline{W_{yW}}^n \\
&+ \frac{1}{\Delta n} R_g \left(\frac{\partial \Delta n}{\partial y} + \frac{\partial N_S}{\partial y} \rho'_{WS} - \frac{\partial N_H}{\partial y} \rho'_{WH} \right) \overline{T_W}^n \\
&\quad + \frac{1}{\Delta n} R_g \left(\frac{\partial N_S}{\partial y} T_{WS} - \frac{\partial N_H}{\partial y} T_{WH} \right) = 0
\end{aligned}
$$

$$
\begin{aligned}
&\frac{\overline{\partial_R T_W}^n}{\partial t}+\overline{W_{mW}}^n\frac{\overline{\partial T_W}^n}{\partial m}+\overline{W_{yW}}^n\frac{\overline{\partial T_W}^n}{\partial y}+(\gamma-1)\overline{T_W}^n\frac{\overline{\partial W_{mW}}^n}{\partial m}\\
&+(\gamma-1)\overline{T_W}^n\frac{\overline{\partial W_{yW}}^n}{\partial y}\\
&\quad+\left[\gamma\frac{1}{\Delta n}\frac{\partial \Delta n}{\partial m}+(\gamma-1)\left(\overline{\left(\frac{\partial\theta}{\partial Y}\right)_W}^n\overline{\sin\varphi_W}^n+\frac{\overline{\partial\varphi_W}^n}{\partial n}\right)\right]\overline{W_{mW}}^n\overline{T_W}^n\\
&+\gamma\frac{1}{\Delta n}\frac{\partial\Delta n}{\partial y}\overline{T_W}^n\overline{W_{yW}}^n\\
&\quad+(\gamma-1)\left(\overline{\left(\frac{\partial\theta}{\partial Y}\right)_W}^n\overline{\cos\varphi_W}^n+\frac{\overline{\partial\varphi_W}^n}{\partial m}\right)\overline{W_{mW}}^n\overline{T_W}^n\\
&+\frac{1}{\Delta n}\left(\frac{\partial N_S}{\partial y}\overline{T_{WS}}^n-\frac{\partial N_H}{\partial y}\overline{T_{WH}}^n\right)\overline{W_{mW}}^n\\
&\quad+\frac{1}{\Delta n}\left(\frac{\partial N_S}{\partial y}\overline{T_{WS}}^n-\frac{\partial N_H}{\partial y}\overline{T_{WH}}^n\right)\overline{W_{yW}}^n+\frac{1}{\Delta n}\left(\overline{T_{WH}}^n-\overline{T_{WS}}^n\right)\overline{W_{nW}}^n\\
&\quad+(\gamma-1)\frac{1}{\Delta n}\left(\begin{aligned}&\overline{W_{nWH}}^n-\overline{W_{nWS}}^n+\frac{\partial N_S}{\partial m}\overline{W_{mWS}}^n-\frac{\partial N_H}{\partial m}\overline{W_{mWH}}^n\\&+\frac{\partial N_S}{\partial y}\overline{W_{yWS}}^n-\frac{\partial N_H}{\partial y}\overline{W_{yWH}}^n\end{aligned}\right)\overline{T_W}^n=0
\end{aligned}
\tag{9.26}
$$

In these equations, the subscript w is replaced by E for the quantities taken on the upper surface and by I for those on the lower surface of the profile of two consecutive blades of the turbine wheel.

The quantities taken from the tips of the blades, hereinafter called wedges, are contained in the equations averaged on the crankcase, hub, upper surface and lower surface.

9.4.3. *Expressions of the equations of motion of flows on the corners*

These are the equations relating to the flows in the channel between two consecutive blades located at the intersections of crankcase: upper surface (HE), hub; upper surface (SE), crankcase; lower surface (HI), hub; lower surface (SI). The equations of motion on these points are given by the following relations:

$$\frac{\partial_R \rho'_{Wb}}{\partial t} + W_{mWb}\frac{\partial \rho'_{Wb}}{\partial m} + W_{yWb}\frac{\partial \rho'_{Wb}}{\partial Y} + W_{nWb}\frac{\partial \rho'_{Wb}}{\partial n}$$
$$+\frac{\partial W_{mWb}}{\partial m} + \frac{\partial W_{yWb}}{\partial Y} + \frac{\partial W_{nWb}}{\partial n}$$
$$+\left[\left(\frac{\partial \theta}{\partial Y}\right)_{Wb} \sin\varphi_{Wb} + \frac{\partial \varphi_{Wb}}{\partial n}\right] W_{mWb} \quad [9.27]$$
$$+\left[\left(\frac{\partial \theta}{\partial Y}\right)_{Wb} \cos\varphi_{Wb} + \frac{\partial \varphi_{Wb}}{\partial m}\right] W_{nWb} = 0$$

$$\frac{\partial_R W_{mWb}}{\partial t} + W_{mWb}\frac{\partial W_{mWb}}{\partial m} + W_{yWb}\frac{\partial W_{mWb}}{\partial y} + W_{nWb}\frac{\partial W_{mWb}}{\partial n}$$
$$+R_g\frac{\partial T_{Wb}}{\partial m} + R_g T_{Wb}\frac{\partial \rho'_{Wb}}{\partial m}$$
$$- W_{mWb}W_{nWb}\frac{\partial \varphi_{Wb}}{\partial m} - W_{nWb}^2\frac{\partial \varphi_{Wb}}{\partial n} - \left(W_{yWb} + \omega R_{Wb}\right)^2 \frac{\sin\varphi_{Wb}}{R_{Wb}} \quad [9.28]$$
$$-W_{yWb}^2\left[\left(\frac{\partial \theta}{\partial Y}\right)_{Wb} R_{Wb} - 1\right]\frac{\sin\varphi_{Wb}}{R_{Wb}} = 0$$

$$\frac{\partial_R W_{nWb}}{\partial t} + W_{mWb}\frac{\partial W_{nWb}}{\partial m} + W_{yWb}\frac{\partial W_{nWb}}{\partial y} + W_{nWb}\frac{\partial W_{nWb}}{\partial n}$$
$$+R_g\frac{\partial T_{Wb}}{\partial n} + R_g T_{Wb}\frac{\partial \rho'_{Wb}}{\partial n}$$
$$+ W_{mWb}^2\frac{\partial \varphi_{Wb}}{\partial m} + W_{mWb}W_{nWb}\frac{\partial \varphi_{Wb}}{\partial n} - \left(W_{yWb} + \omega R_{Wb}\right)^2 \frac{\cos\varphi_{Wb}}{R_{Wb}} \quad [9.29]$$
$$-W_{yWb}^2\left[\left(\frac{\partial \theta}{\partial Y}\right)_{Wb} R_{Wb} - 1\right]\frac{\cos\varphi_{Wb}}{R_{Wb}} = 0$$

$$\frac{\partial_R W_{yWb}}{\partial t} + W_{mWb}\frac{\partial W_{yWb}}{\partial m} + W_{yWb}\frac{\partial W_{yWb}}{\partial y} + W_{nWb}\frac{\partial W_{yWb}}{\partial n} + R_g \frac{\partial T_{Wb}}{\partial y} + R_g T_{Wb}\frac{\partial \rho'_{Wb}}{\partial y} + \left[W_{yWb}\left(\frac{\partial \theta}{\partial Y}\right)_{Wb} R_{Wb} + 2\omega R_{Wb} \right] \left(W_{mWb}\frac{\sin\varphi_{Wb}}{R_{Wb}} + W_{nWb}\frac{\cos\varphi_{Wb}}{R_{Wb}} \right) = 0 \qquad [9.30]$$

$$\frac{\partial_R T_{Wb}}{\partial t} + W_{mWb}\frac{\partial T_{Wb}}{\partial m} + W_{yWb}\frac{\partial T_{Wb}}{\partial y} + W_{nWb}\frac{\partial T_{Wb}}{\partial n} + (\gamma - 1) T_{Wb} \left\{ \begin{array}{l} \dfrac{\partial W_{mWb}}{\partial m} + \dfrac{\partial W_{yWb}}{\partial y} + \dfrac{\partial W_{nWb}}{\partial n} \\ + \left[\left(\dfrac{\partial \theta}{\partial Y}\right)_{Wb} \sin\varphi_{Wb} + \dfrac{\partial \varphi_{Wb}}{\partial n} \right] W_{mWb} \\ + \left[\left(\dfrac{\partial \theta}{\partial Y}\right)_{Wb} \cos\varphi_{Wb} - \dfrac{\partial \varphi_{Wb}}{\partial m} \right] W_{nWb} \end{array} \right\} = 0 \qquad [9.31]$$

In the expressions of the equations of motion on the corners above, the index is replaced by E for the points taken on the upper surface and by I for those on the lower surface, while the index is replaced by H for those of the crankcase and by S for those of the hub.

9.5. Calculating the drive torque of the wheel of a wind turbine

Let $\vec{W}$ be the velocity of a flow passing through a wind turbine consisting of a wheel in a volume Ω bounded by a surface S.

The boundaries composing the surface S delimiting the volume Ω are formed by the fluid surface boundaries $(S_l) = (S_e) + (S_S) + (S_C)$ and solid walls (S_W) which correspond to the contour surfaces of the wheel, hub and possibly crankcase as shown in Figure 9.5.

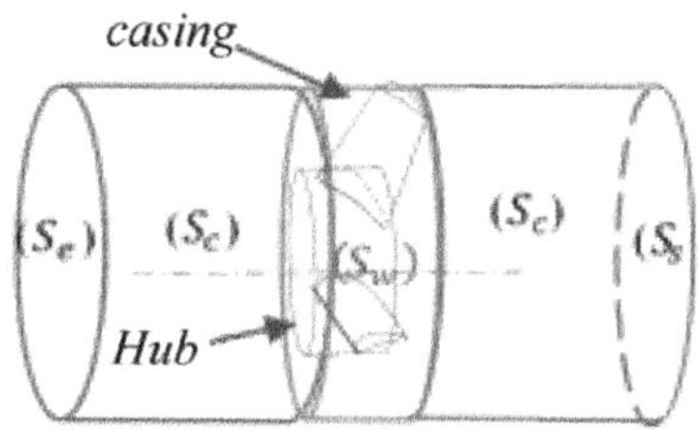

Figure 9.5. *Representative diagram of the volume Ω*

The temporal variation of momentum $\frac{d\left(m\vec{W}\right)}{dt}$ of the fluid is due to the sum of the forces that are applied to the fluids in accordance with Newton's second law. These forces are composed of pressure forces, viscosity forces, weight, Coriolis force, and drive force.

Where from:

$$\frac{d(m\vec{W})}{dt} = \sum \vec{F}_{applied} \qquad [9.32]$$

thus:

$$\frac{\partial_R}{\partial t}\iiint_{\Omega} dm\vec{W} + \iint_S dm_S\vec{W} = \iint_S \left(-p\vec{n} - \vec{\zeta}\,\tau\right)ds - \iiint_{\Omega}\rho g dv - 2\rho\iiint_{\Omega}\left(\vec{\omega}\wedge\vec{W}\right)dv - \rho\iiint_{\Omega}\left(\vec{\omega}\wedge(\vec{\omega}\wedge\vec{r})\right)dv \qquad [9.33]$$

In equation [9.33], we have $dm = \rho dv$ and $dm_S = \rho\vec{W}\,\vec{n}dv$, $\vec{\zeta}$ is a unit vector tangent to the elementary ds surface taken on s whose normal vector $\vec{n}$, τ is the component of the viscosity forces following. The expansion of equation [9.33] gives the following equation $\vec{\zeta}$:

$$\begin{aligned}\frac{\partial_R}{\partial t}\iiint_{\Omega} dm\vec{W} - \iint_{Ss} dm_{Ss}\vec{W}_s + \iint_{Se} dm_{Se}\vec{W}_e &= \iint_{Ss}\left(-\vec{n}_s p_s\, ds_s\right) + \iint_{Ss}\left(-\vec{i}_s\tau_s\, ds_s\right)\\ &+ \iint_{Se}\left(-\vec{n}_e p_e\, ds_e\right) + \iint_{Se}\left(-\vec{i}_e\tau_e\, ds_e\right) - \left[\iint_{Sw}\left(\vec{n}_w p_w\, ds_w\right) + \iint_{Sw}\left(-\vec{i}_w\tau_w\, ds_w\right)\right]\\ &- \iiint_{\Omega} 2\rho\vec{\omega}\wedge\vec{W}dv - \iint_{\Omega}\rho\vec{\omega}\wedge(\vec{\omega}\wedge\vec{r})dv - \iiint_{\Omega}\rho\vec{g}dv\end{aligned} \qquad [9.34]$$

$\vec{R}$ is the resultant of the forces applied to the wheel bounded by (S_W) and is equal to:

$$\vec{R} = \iint_{Sw} \vec{n}_w p_w \, ds_w + \iint_{Sw} \vec{i}_w \tau_w \, ds_w \quad [9.35]$$

The projection $\vec{R}_a$ of $\vec{R}$ on the direction of the wheel axle has the expression:

$$\vec{R}_a = \iint_{S_e} dms_e \vec{i} Wa_e - \iint_{S_s} dms_s \vec{i} Wa_s + \iint_{S_e} -\vec{i} p_e ds_e \cos\lambda_e - \iint_{S_s} -\vec{i} p_s ds_s \cos\lambda_s \quad [9.36]$$

λ is the angle that the axis of rotation makes with the direction of flow velocity as in the Figure 9.6.

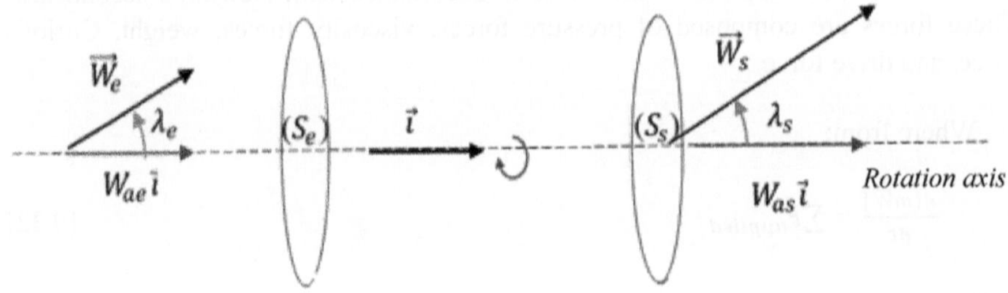

Figure 9.6. *Compute domain input and output surface*

The equation of momentum in an elementary volume taken from the volume is written as follows:

$$\frac{\partial(\rho\vec{V})}{\partial t} + \vec{\nabla}\cdot(\rho\vec{V} \otimes \vec{V}) = f_s + \rho\vec{g} \quad [9.37]$$

f_s is the surface force.

Form [9.37] can be written as follows:

$$\frac{\partial}{\partial t}\iiint_\Omega \vec{r} \wedge \rho\vec{V} dv + \iiint_\Omega \vec{r} \times \vec{\nabla}\cdot(\rho\vec{V} \otimes \vec{V}) dv = \iint_S \vec{r} \wedge (-\vec{n} p ds - \vec{i}\tau ds) + \iiint_\Omega \rho\vec{r} \wedge \vec{g} dv \quad [9.38]$$

thus:

$$\frac{\partial}{\partial t}\iiint_{\Omega}\vec{r}\wedge\vec{V}dm+\iint_{S}\vec{r}\wedge\vec{V}dm=\iint_{S}\vec{r}\wedge\left(-\vec{n}p-\vec{i}\,\tau\right)ds+\iiint_{\Omega}\vec{r}\wedge\vec{g}dm \quad [9.39]$$

In the steady state, the moment of forces applied to the bounded wheel (S_W) is expressed as follows:

$$\overrightarrow{M_0}=\iint_{S}\vec{r}_w\wedge\left(-\vec{n}_w p_w-\vec{i}_w\tau_w\right)ds_w \quad [9.40]$$

thus:

$$\begin{aligned}\overrightarrow{M_0}=&-\iiint_{\Omega}\vec{r}\wedge\rho\vec{V}dv-\iint_{Ss}\vec{r}_s\wedge\vec{V}_s dm_{Ss}+\iint_{Se}\vec{r}_e\wedge\vec{V}_e dm_{Se}+\iiint_{\Omega}\vec{r}\wedge\vec{g}dm\\&+\iint_{Ss}\left(-\vec{r}_s\wedge\vec{n}_s p_s ds_S\right)\\&+\iint_{Ss}\left(-\vec{r}_s\wedge\vec{i}_s\tau_s ds_S\right)+\iint_{Se}\left(-\vec{r}_e\wedge\vec{n}_e p_e ds_e\right)\\&+\iint_{Se}\left(-\vec{r}_e\wedge\vec{i}_e\tau_e ds_e\right)\end{aligned} \quad [9.41]$$

Replacing $\vec{V}$ in equation [9.41] by $\vec{W}+\vec{\omega}\wedge\vec{r}$, neglecting the weight and considering that the flow is stationary, the moment of the forces applied to the wheel with respect to the axis of rotation $\left(\vec{i}\right)$ is:

$$M_a=\vec{i}\,\overrightarrow{M_0} \quad [9.42]$$

Replacing $\overrightarrow{M_0}$ by its expression [9.41] and after simplification, we obtain:

$$\overrightarrow{M_a}=-\iint_{Ss}\left(R_s W_{us}+\omega R_s^2\right)dm_{Ss}+\iint_{Se}\left(R_e W_{ue}+\omega R_e^2\right)dm_{Se} \quad [9.43]$$

Thus, the torque applied by the average flow of the fluid on the impeller has the following expression:

$$\vec{M}_{average}=\left[\overline{\overline{\rho_e Q_e}}\left(\overline{\overline{W}}_{ue}+\omega\overline{\overline{R}}_e\right)\overline{\overline{R}}_e-\overline{\overline{\rho_s Q_s}}\left(\overline{\overline{W}}_{us}+\omega\overline{\overline{R}}_s\right)\overline{\overline{R}}_s\right]\frac{\vec{\omega}}{\omega} \quad [9.44]$$

$\overline{\overline{Q}}_e, \overline{\overline{W}}_{ue}, \overline{\overline{R}}_e$ and $\overline{\overline{Q}}_s, \overline{\overline{W}}_{us}, \overline{\overline{R}}_s$ are, respectively, the mean volume flow, the tangential component of the mean relative velocities and the average radii at the inlet and outlet of the impeller.

9.6. Different power expression and efficiency calculation

9.6.1. *Available power*

The available power corresponds to the power coming from the kinetic energy of the fluid at the site where the machine is placed. It depends essentially on the density ρ of the fluid, the cross section S of the sensor and the speed V of attack of the wind. It is expressed by the following relationship:

$$P_{available} = \frac{1}{2} \rho S V^3 \quad [9.45]$$

9.6.2. *Starting power*

The inertia of the system and friction contribute to the immobilization of the wind turbine. In order for the shaft of the latter to rotate, the available power must be greater than the starting power corresponding to a minimum speed V_D of the site's fluid:

$$P_D = \frac{1}{2} \rho S V_D^3 \quad [9.46]$$

9.6.3. *Mechanical power*

The mechanical power recovered from the shaft of the fluid machine is the product of the torque applied to the shaft by its angular velocity. For an average flow causing an average torque $M_{average}$ to the shaft rotating at an angular velocity ω, this power is:

$$P_{average} = M_{average}\omega = \left[\overline{\overline{\rho_S Q_S}}\left(W_{uS} + \omega \overset{=}{R}_S\right)\overset{=}{R}_S - \overline{\overline{\rho_e Q_e}}\left(W_{ue} + \omega \overset{=}{R}_e\right)\overset{=}{R}_e\right]\omega \quad [9.47]$$

9.6.4. *Performance*

Yield η of the conversion of the wind turbine is the quotient of the average mechanical power by the available power. It is expressed as:

$$\eta = \frac{P_{average}}{P_{available}} \quad [9.48]$$

9.7. Numerical resolution method used

9.7.1. *Mesh*

The mesh used for numerical simulation is an orthogonal mesh with irregular steps whose boundaries are represented by Figure 9.7.

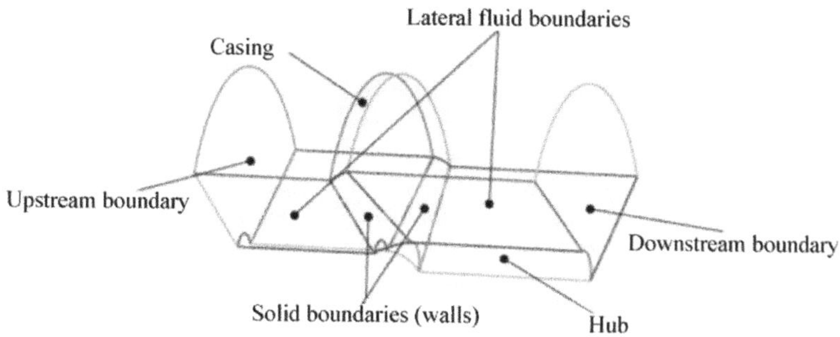

Figure 9.7. *Presentation of the boundary of the field of computing*

9.7.2. *Discretization scheme*

The discretization is carried out according to two schemes:

– spatial discretization scheme;

– temporal discretization schemes.

9.7.3. *Spatial discretization scheme*

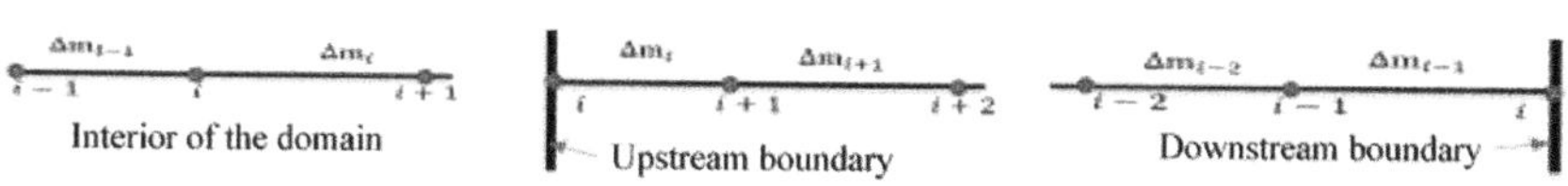

Figure 9.8. *Presentation of the mesh index i node with its irregular steps*

9.7.3.1. *Time step presentation*

The temporal discretization scheme adopted is the explicit scheme

$$\frac{\partial F}{\partial t} = \frac{F^{t} - F^{t+\Delta t}}{\Delta t} \quad [9.49]$$

Node location	Discretization schemes of $\frac{\partial F}{\partial m}$
Inside the computational domain	$\frac{\partial F(i)}{\partial m}=\frac{\Delta m_{i-1}}{\Delta m_i\left(\Delta m_i+\Delta m_{i-1}\right)}F(i+1)-\frac{\Delta m_{i-1}-\Delta m_i}{\Delta m_i \Delta m_{i-1}}F(i)-\frac{\Delta m_i}{\Delta m_{i-1}\left(\Delta m_i+\Delta m_{i-1}\right)}F(i-1)$
Upstream boundary	$\frac{\partial F(i)}{\partial m}=\frac{\Delta m_i}{\Delta m_{i+1}\left(\Delta m_i+\Delta m_{i+1}\right)}F(i+2)+\frac{2\Delta m_{i+1}+\Delta m_i}{\Delta m_i \Delta m_{i+1}}F(i)+\frac{\Delta m_i+\Delta m_{i+1}}{\left(\Delta m_i+\Delta m_{i-1}\right)}F(i+1)$
Downstream boundary	$\frac{\partial F(i)}{\partial m}=\frac{\Delta m_{i-1}}{\Delta m_{i-2}\left(\Delta m_{i-1}+\Delta m_{i-2}\right)}F(i-2)+\frac{2\Delta m_{i-1}+\Delta m_{i-2}}{\Delta m_{i-1}\left(\Delta m_{i-2}+\Delta m_{i-1}\right)}F(i)-\frac{\Delta m_{i-1}+\Delta m_{i-2}}{\left(\Delta m_{i-1}\Delta m_{i-2}\right)}F(i-1)$

Table 9.3. *Discretization schemes for the derivative with respect to m of a function F*

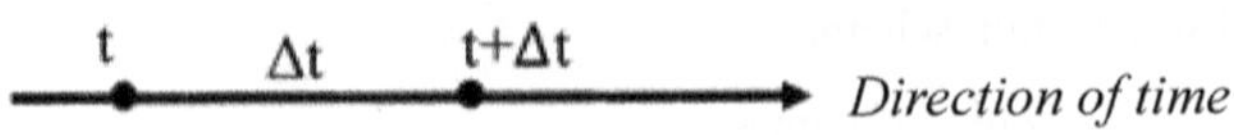

Figure 9.9. *Presentation of the time step*

9.7.4. *Synthesis*

9.7.4.1. *Summary*

At time t, at any node of index i of the mesh, the systems of equations of motion [9.12]–[9.15], [9.17]–[9.21], [9.22]–[9.26] and [9.27]–[9.31] can be summarized as follows: $\frac{\partial\left[\vec{Z}(i)\right]^t}{\partial t} = \left[\overrightarrow{RHS}(i)\right]^t$ with $\left[\vec{Z}(i)\right]^t = \begin{pmatrix} \rho' \\ W_m \\ W_n \\ W_y \\ T \end{pmatrix}_i^t$ and i is the index of the next node (m). [9.50]

The numerical difference method

Finite is an explicit two-step time method with a two-order precision (spatial and temporal).

Characteristic flow variables are calculated at time t and $t+\Delta t$. Either t^* a time between t and $t+\Delta t$.

– First step: the prediction step:

This step calculates the characteristic quantities of the flow at the node of index i at the time t^* $t^* = t + \Delta t^*$ according to the following discretization scheme:

$$\left[\vec{Z}(i)\right]^{t^*} = \frac{\left(\left[\vec{Z}(i+1)\right]^t + \left[\vec{Z}(i)\right]^t + \left[\vec{Z}(i-1)\right]^t\right)}{3} + \Delta t^*\left[\overrightarrow{RHS}(i)\right]^t \quad [9.51]$$

– Second step: the correction step:

This allows us to calculate the characteristic quantities of the flow at the node of index i at the instant $t_2 = t^* + \Delta t_2$. At the instant t_2, the temporal discretization of a quantity $\vec{Z}(i)$ It will be done in the following way:

$$\left[\vec{Z}(i)\right]^{t_2} + \left[\vec{Z}(i)\right]^t + \Delta t_2\left[\overrightarrow{RHS}(i)\right]^{t^*} \quad [9.52]$$

The internal stability condition of the numerical calculation schemes adopted is ensured by the CFL (Friedrich Lévy Current) condition:

$$\Delta t^* < \frac{l(i)}{\sqrt{3}(W^t+a^t)_{(i)}} \ and \ \Delta t_2 < \frac{l(i)}{\sqrt{3}(W^*+a^*)_{(i)}} \quad [9.53]$$

with W being the modulus of the relative velocity and a being the modulus of the speed of sound. Quantities whose exponent is * correspond to the quantities calculated at the intermediate time t^*. $l(i)$ is the smallest distance between two neighboring nodes of the mesh, W^t, a^t, $W^*$and a^* the greatest of the relative speeds and the speeds of sound at the instant t and t^* in the field of computing.

9.7.4.2. *Initial conditions used*

Following a mesh line assumed to be confused with a streamline.

The flow rate remains constant in all normal sections A_i at $\vec{m}$. The result is:

$$\rho_i A_i W_{mi} = const \qquad [9.54]$$

Following a mesh line assumed to be confused with a streamline:

– Rothalpy remains constant:

$$I = \frac{W^2}{2} C_P T + \frac{\omega^2 r^2}{2} = const \qquad [9.55]$$

– Entropy is assumed to be constant:

$$\frac{\rho}{\rho_1} = \left(\frac{T}{T_1}\right)^{\frac{1}{\gamma - 1}} \qquad [9.56]$$

9.7.4.3. *Boundary conditions applied*

The boundary conditions are summarized in Table 9.4. They are imposed on the borders after the realizations, the prediction stage and the correction stage.

Upstream boundary	**Downstream boundary**	**On solid side walls**	**On lateral fluid borders**	**Casing**	**Hub**
Speed $V_0 =$ 14 m/s And angle $\alpha_0 = 27°$	Static pressure = 101325. Pascal	Sliding condition: $W_y = W_m$ tan $\beta + W_n \tan \delta$	Periodicity condition (equality of values of quantities on both boundaries)	Static pressure = 101325. Pascal	Non-penetration condition $W_n = 0$

Table 9.4. *Boundary conditions adopted*

9.7.4.4. *Convergence test*

The calculation converges when the values of the density at any point in the computational domain at the Nth iteration are different by ε from their values at the Mth iteration. With M = N + NCO. This translates into:

$$\Delta RO = \left|(\rho)^{N} - (\rho)^{N+NCO}\right| \leq \varepsilon \qquad [9.57]$$

with ε is in the 10^{-5} range

9.8. Application of the method

9.8.1. *System description*

The method is applied to a three-bladed wind turbine whose geometry is shown in Figure 9.10, giving the details in Figures 9.11–9.14.

Figure 9.10. *Three-bladed light*

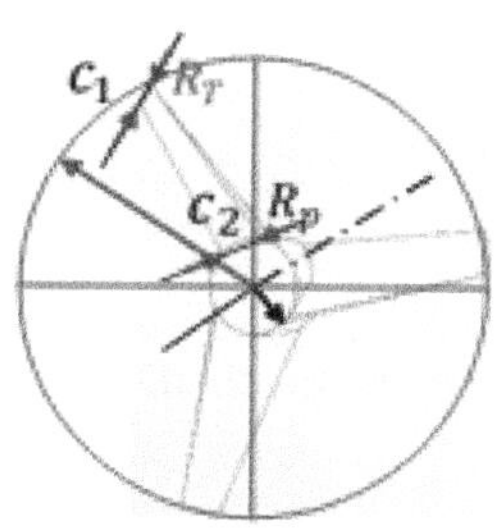

Figure 9.11. *Wheel geometry*

Figure 9.12. *Inter-blade channel Head radius $R_T = 0.8$ m, foot radius $R_p = 0.02m$, reduction report $\frac{c_1}{c_2} = \frac{1}{3}$ and cushioning = 27°*

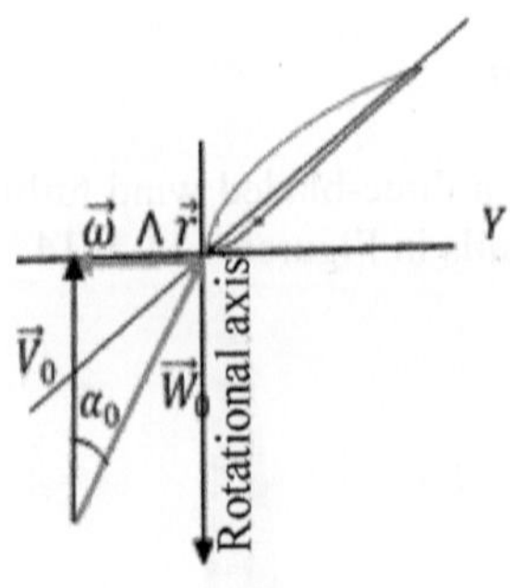

Figure 9.13. *Upstream boundary velocity*

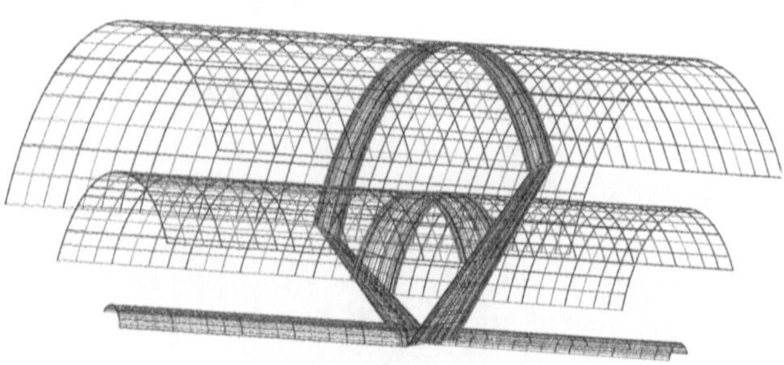

Figure 9.14. *Appearance of the mesh 3D in R_p, at mid-radius and in R_T*

Figure 9.14 shows the 3D meshing of the computational domains.

9.8.2. *Calculation process*

The calculation is made according to the organizational chart in Figure 9.15.

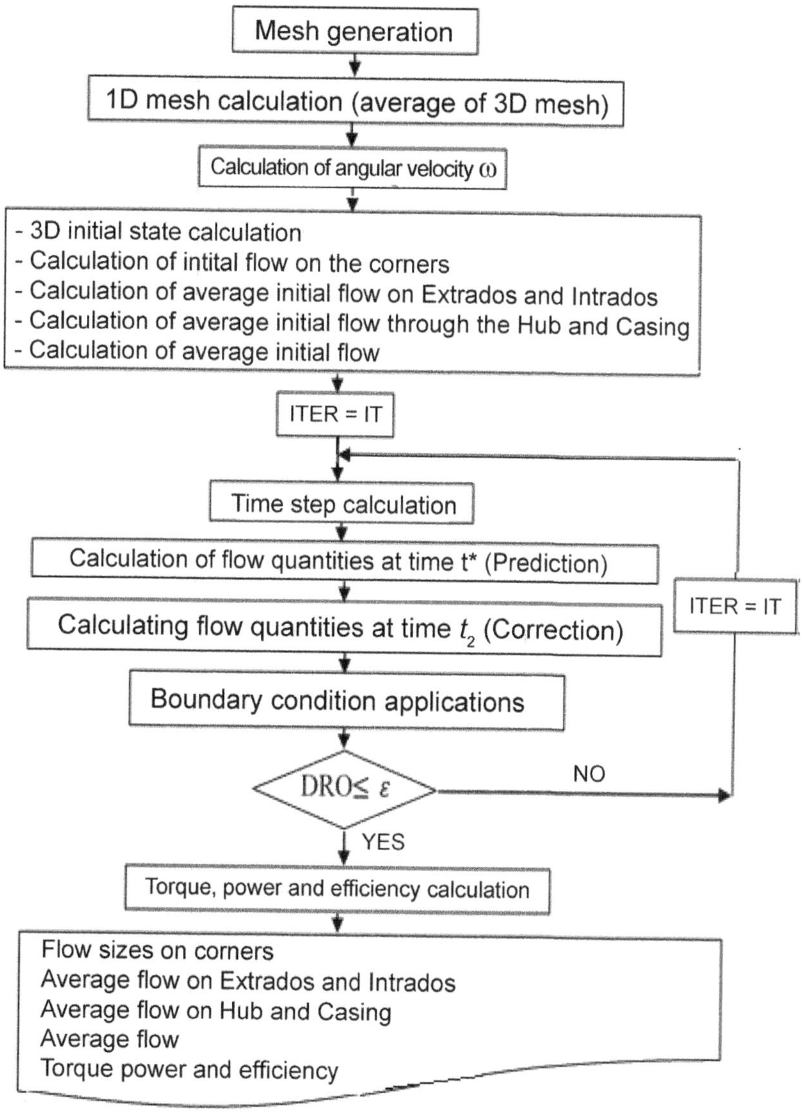

Figure 9.15. *Flowchart of the performance calculation process of a wind turbine*

9.9. Results and interpretation

9.9.1. *Initial flow velocity within the inter-blade channel*

The velocity of the flow in the initial state inside the inter-blade channel of the machine is shown in Figure 9.16.

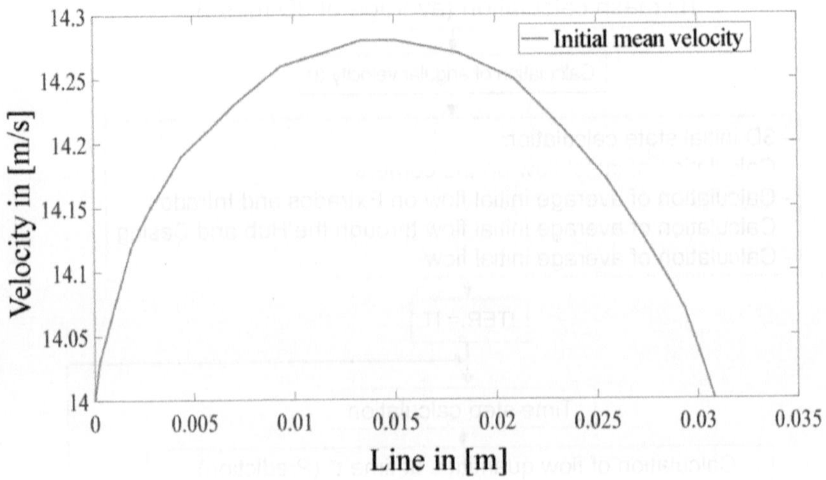

Figure 9.16. *Evolution of the initial mean velocity in the inter-blade channel*

9.9.2. *Convergence of calculations based on NCO*

The curves of evolution of $\Delta RO = \left|(\rho)^N - (\rho)^{N+NCO}\right|$ according to the number of iterations are shown in Figure 9.10. ΔRO is calculated from the density of the average flow.

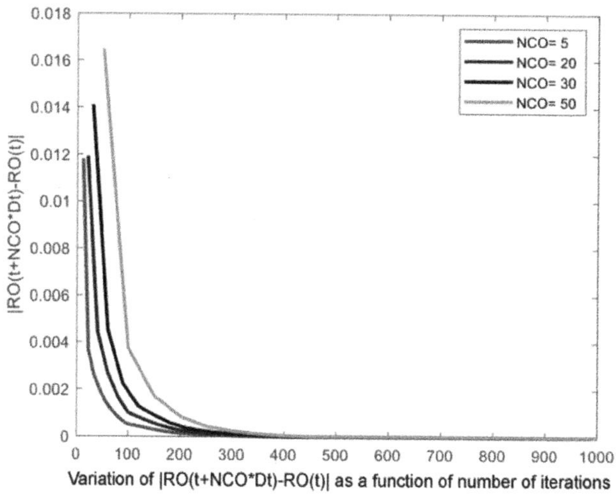

Figure 9.17. *Presentation of convergence curves for various NCO values NCO is taken equal to 50 for the following calculations*

9.9.3. *Performances*

9.9.3.1. *Torque*

The curve representing the torque applied by the wind to the wind turbine wheel as a function of the upstream speed of the wheel is shown in Figure 9.18.

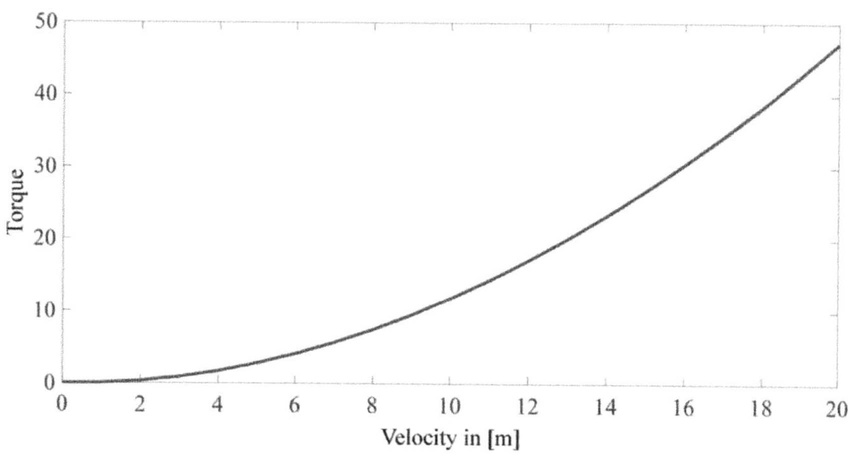

Figure 9.18. *Torque produced by the averaged flow of air through the impeller*

9.9.3.2. *Powers*

The available power and the power collected at the wheel shaft as a function of wind speed are shown in Figure 9.19.

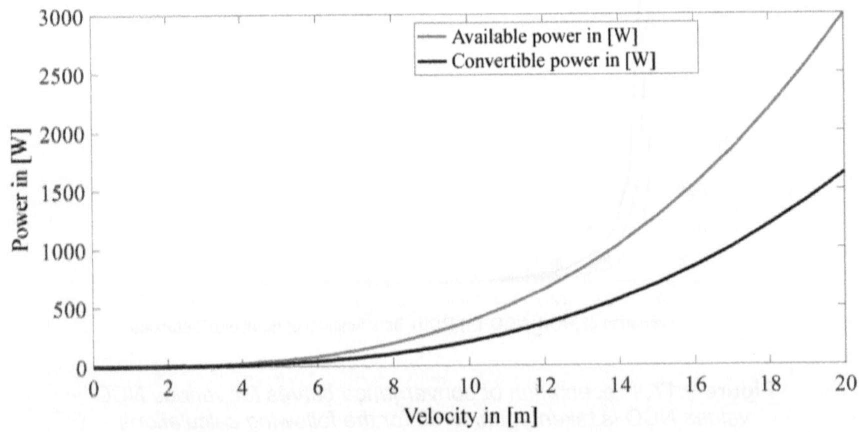

Figure 9.19. *Available and convertible power supply*

9.9.3.3. *Performance*

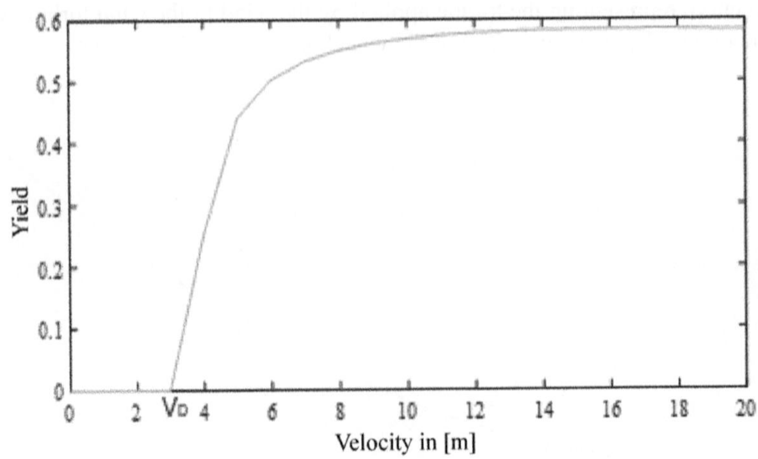

Figure 9.20. *Performance curve of the conversion of this approach*

The shape of the yield curve in formula [9.46] as a function of wind speed is given in Figure 9.20. As long as the machine is not running, the power convertible into electrical power is zero.

9.9.4. *Interpretation of the results obtained*

The simulation results show that torque remains low for upstream wind speeds below 5 m/s. The available power and the convertible power at the wheel shaft remain very close for speeds below 5 m/s and above 5 m/s. The difference is particularly noticeable for speeds above 10 m/s. As for the yield, it increases with the upstream speed and tends asymptotically towards a value of 58.40%. The comparison of the results obtained by the numerical method in question with those of the methods recommended by Albert Betz and Gorban and Gorlov, applied to a wind turbine with a horizontal axis, is given in Table 9.5. If Betz, in his approach, considers that the flow is one-dimensional and that the impeller is assumed to be an infinitely thin disk per fluid that flows at a certain speed, Gorban and Gorlov have taken into account the tangential component of the flow velocity.

Approaches	**Betz**	**Gorban and Gorlov**	**This chapter**
Performance η	59.3%	30.1%	58.40%

Table 9.5. *Comparison between the limits of Betz, Gorban and Gorlov and that resulting from the method applied to a horizontal axis wind turbine*

9.10. Conclusion and perspective

This chapter has been devoted to the search for a one-dimensional model equivalent to a three-dimensional flow whose solution of the associated systems of nonlinear partial differential equations requires a lot of time [BEN 08; BEA 08; BEA 14]. To this end, they proceeded to calculate the averages of the characteristic quantities of the fluid flow through the sensor of a turbomachine. They obtained a one-dimensional system that is lighter in terms of calculation.

The simulations showed:

– satisfactory results, in particular a yield between Betz and Gorban and Gorlov;

– a low computation time compared to the time required by three-dimensional numerical simulations;

– the effectiveness of the method applied to a horizontal-axis wind turbine which allows the authors to propose it as a design tool for horizontal-axis machines.

Its confirmation requires its application to vertical-axis machines. This continuation will contribute to its improvement.

9.11. References

[AIT 17] AIT MOHAMMED M., "Étude des Tuyères composites pour une conception optimale d'une hydrolienne à axe horizontal" PhD Thesis, Université de Bretagne Occidentale, Brest, 2017.

[ASU 03] ASUAJE M., Méthodologie et optimisation dans la conception et l'analyse des performances des turbomachines a fluide incompressible, PhD Thesis, École Nationale Supérieure d'Arts et Métiers Centre de Paris, Paris, 2003.

[BEA 08] BEAUME G., Modélisation et simulation numérique de l'écoulement d'un fluide complexe, PhD Thesis, École Nationale Supérieure des Mines de Paris, Paris, available at: https://pastel.archives-ouvertes.fr/tel-00416435, 2008.

[BEA 14] BEAUDET L., Étude expérimentale et numérique du décrochage dynamique sur une éolienne à axe vertical de forte solidité, PhD Thesis, Université de Poitiers, Poitiers, available at: https://tel.archives-ouvertes.fr/tel-01058929, 2014.

[BEN 08] BEN HAJ ALI A., Calcul de haute performance en aéroélasticité et en écoulements turbulents tridimensionnels, PhD Thesis, École de technologie supérieure, Université du Québec, Quebec, 2008.

[BOU 16] BOUMRAR I., "Effets de l'angle de calage des pales d'une petite eolienne sur sa puissance electrique generee", in *Third International Conférence on Energy, Materials, Applied Energetics and Pollution, ICEMAEP2016*, Constantine, 2016.

[CHA 13] CHATELIN R., Méthodes numériques pour l'écoulement de Stokes 3D : fluides à viscosité variable en géométrie complexe mobile ; application aux fluides biologiques, PhD Thesis, Université Toulouse III Paul Sabatier, Toulouse, available at: https://tel.archives-ouvertes.fr/tel-00946993, 2013.

[DAG 13] DAGUSE B., Modélisation analytique pour le dimensionnement par optimisation d'une machine dédiée à une chaîne de traction hybride à dominante électrique, PhD Thesis, CentraleSupélec, Gif-sur-Yvette, 2013.

[ELM 19] EL MAANI R., RADI B., EL HAMI A., "Simulation numérique de l'écoulement autour d'une éolienne à axe horizontal", *Incertitudes et fiabilité des systèmes multiphysiques*, 3, doi: 10.21494/ISTE.OP.2019.0432, 2019.

[HAM 10] HAMMOUTI A., Simulation numérique directe en différence finie de l'écoulement d'un fluide incompressible en présence d'interfaces rigides, PhD Thesis, École Nationale des Ponts et Chaussées, Champs-sur-Marn, available at: https://pastel.archives-ouvertes.fr/pastel-00511607, 2010.

[PEN 11] PENIN V., Interactions rotor-stator en turbine : étude de l'effet potentiel remontant, PhD Thesis, École Centrale de Lyon, Lyon, available at: https://tel.archives-ouvertes.fr/tel-00679727, 2011.

[RAJ 23] RAJOHARISOA L.A., RANDRIATEFISON N., “Éolienne artisanale pour l’éducation sur l’énergie renouvelable”, *Entropie thermodynamique – énergie – environnement – économie*, 4(1), doi: 10.21494/ISTE.OP.2023.0941, 2023.

[RAV 19] RAVELOSON F., TOMBORAVO D., VONY R., “Étude de l’écoulement tridimensionnel et instationnaire dans un rotor d’une éolienne Savonius”, *Thermodynamique des interfaces et mécanique des fluides*, 3(1), doi: 10.21494/ISTE.OP.2019.0357, 2019.

[SAL 16] SALAMEH G., Caractérisation expérimentale d’une turbine de suralimentation automobile et modélisation de ses courbes caractéristiques de fonctionnement, PhD Thesis, École Centrale de Nantes de l’Université Bretagne Loire, Nantes, available at: https://hal.archives-ouvertes.fr/tel-01635671, 2016.

[SMI 20] SMILY J., RADI B., SOSSEY-ALAOUI I., “Simulation numérique et validation expérimentale de l’écoulement d’air autour d’un profil de pale d’éolienne de type S809”, *Incertitudes et fiabilité des systèmes multiphysiques*, 4(1), doi: 10.21494/ISTE.OP.2020.0566, 2020.

[SMI 22] SMILY J., RADI B., SOSSEY-ALAOUI I., “Analyse de fiabilité d'une pale d’éolienne”, *Incertitudes et fiabilité des systèmes multiphysiques*, 6(2), doi: 10.21494/ISTE.OP.2022.0882, 2022

[SOS 18] SOSSEY-ALAOUI I., RADI B., “Une étude aérodynamique d’une pale d’éolienne à axe horizontal”, *Incertitudes et fiabilité des systèmes multiphysiques*, 2(1), doi: 10.21494/ISTE.OP.2018.0257, 2018.

[TOM 22] TOMBORAVO D., SIAKA E.F., VELONTSOA J. et al., “Contrainte électrique majeure d’une génératrice synchrone à rotor bobiné associée à un aérogénérateur chargé et influence de rayon de pales sur la vitesse du vent au démarrage du groupe”, *Entropie thermodynamique – énergie – environnement – économie*, 3(2), doi: 10.21494/ISTE.OP.2022.0863, 2022.

10

Circulation in the Vessels of the Human Body

Blood circulation has fascinated minds since the Egyptians, but it was not until the 17th century and William Harvey that a coherent view was formed, at the same time as a real revolution in medicine. After a brief history, a precise seemingly indispensable description of the heart (the motor of movement) is given, followed by that of the multiple and varied vessels in which the blood flows and exchanges with the organs. This is followed by a section on fluid mechanics and an incentive conclusion.

List of symbols

A, A_0 m^2	Cross-sectional area, reference area
$\vec{\vec{\mathbf{D}}}$ s^{-1}	Strain rate tensor
E Pa	Young's modulus
$\vec{f}$	N.m^{-3} Ranged force
h J.kg^{-1}, m	Enthalpy, arterial thickness
p Pa	Thermodynamic pressure
$\vec{\vec{\mathbf{P}}}$ Pa	Pressure tensor

Chapter written by Roger PRUD'HOMME, François BOUSTANI and Stéphane VINCENT.

Q $m^3.t^{-1}$	Volume flow rate
$\vec{\mathbf{q}}$ $W.m^{-2}$	Heat flow
$R = 8{,}314$ $J.mol^{-1}.K^{-1}$, R m	Gas constant, Radius of a pipe
S $J.K^{-1}$	Entropy
T K	Absolute temperature
t s	Time
$\vec{\mathbf{V}}$, $\vec{\mathbf{U}}$ $m.s^{-1}$	Velocity vector
V $N.m^{-2}$	Force potential at a distance
x, y, z m	Cartesian coordinates
k $N.m^{-2}.s$	Volume viscosity
Δl m	Pipe length
λ $W.m^{-1}.K^{-1}$	Thermal conductivity
ϕ $m^2.t$-^{1}	Velocity potential, viscoelastic coefficient
μ $kg.m^{-1}.s^{-1}$	Shear viscosity
ρ $kg.m^{-3}$	Density
Σ N.m-1	Surface tension
π Pa	Average normal pressure
Λ without dim.	Pressure loss coefficient
v *without im.* $m^2.s^{-1}$	Poisson's ratio, Kinematic viscosity
e $J.kg^{-1}$	Internal energy per unit mass

r, z m.s^{-1}	Cylindrical coordinates
u, v, w m.s^{-1}	Cartesian velocity components
Ur, U_θ, U_z $m.s^{-1}$	Cylindrical speed components
V m^3	Volume
$\dot{W}_s$ Pa.K^{-1}.s^{-1}	Entropy production rate

10.1. Introduction

The fluids of the human body are multiple: physiological serum, chyle, blood, to name a few. Our interest will mainly focus on blood, which is a multiphasic medium, irrigating the various areas: heart, muscles, lungs, brain. The heart plays a very particular role as the engine of blood circulation. Its functions have not always been well accepted, which indicates a history of blood circulation. Blood can sometimes be considered as a continuous medium. It depends on the sizes of its heterogeneities compared to those of the conduits concerned. The vessels are mainly arteries, veins and capillaries. Their sizes and structures are different. The walls of these conduits are in permanent interaction with the blood.

Classical fluid mechanics is of obvious interest when it comes to relating blood pressure to circulation velocity and various pathologies. But things quickly become complicated when we consider the complex behavior of the vessels themselves, and the multiphase environment in which they circulate (Comolet 1984).

10.2. Blood circulation

10.2.1. *A brief history of blood circulation*

The history of the discovery of blood circulation in the human body goes back several centuries since the Egyptians, with exchanges between East and West.

It is described by several authors. A description can be found in Wikipedia[1]. One of the best narratives is found in a work by Boustani (2007).

1 Circulatory system: https://en.wikipedia.org/wiki/Circulatory_system.

There are two steps in its history:

1) The doctrine of Claudius Galenus (Ancient Greek physician, 129–211), marked by the concepts of the ancient Greeks and Aristotle. Galenus brought new knowledge to his time because he "will definitively admit the presence of blood in the arteries, but on the other hand invented a fanciful passage through the interventricular septum – a major anatomical error, which will continue for fourteen centuries in the West!". Galenus' conception appears clearly in Figure 10.1.

2) The brilliant discovery of William Harvey (English physician and physiologist, 1578–1667), based on the discovery of pulmonary circulation in the 16th century, which led to the modern conception of blood circulation shown in Figure 10.2. Harvey's scientific method: Harvey was the first to introduce the quantitative method into medicine[2].

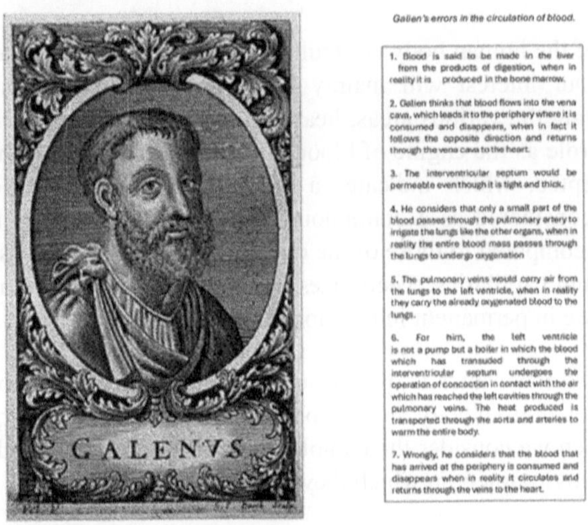

Figure 10.1. *Engraving by Georg Paul Busch[3]. Blood circulation is seen at the beginning of our era[4]*

The blood that has irrigated the body goes back up through the vena cava towards the heart and enters it through the right atrium. It progresses through a valve in the right ventricle which, by contracting, ejects it towards the lungs through the pulmonary artery.

2 The simultaneity of these two aspects of the revolution in medicine, discovery and measurements, has something similar to the revolution in chemistry carried out by Lavoisier nearly two centuries later: discovery and quantitative measurements.

3 From Claude Galenus (see: https://en.wikipedia.org/wiki/Galen).

4 Diagram taken from Boustani's book (2007).

There, it exchanges its CO_2 for oxygen through a network of capillaries and the pulmonary alveoli, then it is returned to the heart – in the left atrium – by the pulmonary veins. This oxygenated blood passes, crossing a valve, into the left ventricle which, by contracting, ejects it into the body (organs, muscles, brain), which it irrigates during exchanges through networks of capillaries and finally comes out deoxygenated. Note that the chemical analysis (Prud'homme et al. 2023) was only possible much later[5].

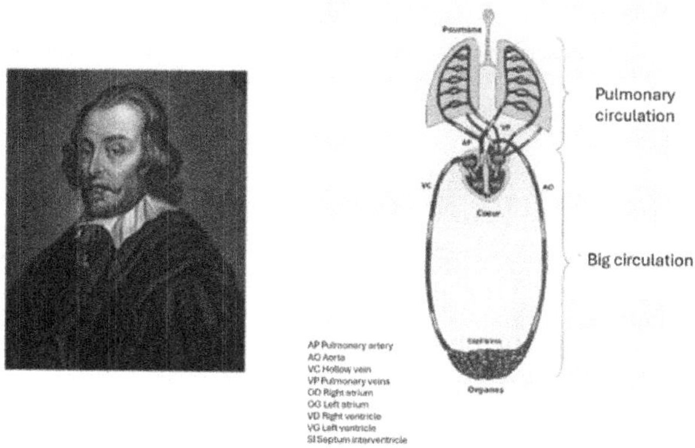

Figure 10.2. *Portrait of William Harvey[6]. Updated view of blood circulation[7]. For a color version of this figure, see www.iste.co.uk/prudhomme/fluid3.zip*

This well-synchronized process occurs approximately once per second throughout life. A more detailed history would highlight that this revolution in concepts was the result of the work and reflections of many scientists[8]. Box 10.1 gives the dates and origins of the scholars who have left their mark on the field.

---

5 Let us note, on the one hand, that Harvey was not able to observe the capillaries himself. Indeed, this observation was only possible later, in 1661, thanks to the invention of the microscope. On the other hand, oxygen and CO_2 were not known at the time of Harvey and it was from the end of the 18th century, with Antoine Lavoisier, that chemical analysis made it possible to specify the nature of the exchanges at the level of the capillaries. Today, measuring the blood oxygenation rate has become commonplace in hospitals.

6 Excerpt from William Harvey (1578–1657) (see: medarus.org).

7 Diagram taken from the book by Boustani (2007).

8 "All the discoveries, prior to Harvey, do not diminish the glory of this great man. There are very few discoveries springing whole from the brain of a single man, as Minerva came fully armed from the brow of Jupiter. They are prepared, matured, anticipated for a long time; then a man comes along who gathers together the scattered facts, takes up, discusses, clears up the confused ideas of his unconscious predecessors, and finally establishes the truth. Such was

The *Egyptians* identified blood as the source of life and the seat of the soul.

Ancient Greece:

– Aristotle (384–322 BCE, ancient Greek philosopher and *polymath*).

– Hippocrates (460–317 BCE, father of medicine, philosopher).

– Herophilos (335–280 BCE, doctor, palpation of the pulse).

– Erasistratus (320–250 BCE, physician, anatomist, venous valves).

– Galenus (131–201, describes the network of veins and arteries from the dissection of pigs).

Muslim doctors:

– Avicenna (Persia, 980–1037, physician philosopher, compatibility of medicine with monotheism).

– Avenzoar (Seville, 1073–1162, experimental physician).

– Ibn Al-Nafis (Syria, 1210–1288, pulmonary circulation, coronary arteries, capillary circulation).

Europe:

– Andreas Vesalius (Belgium, 1514–1564, adapts Galenus' physiological theory to his new observations).

– Amatus Lusitanus (Spain and Portugal, 1511–1568, veins have valved that force blood back to the heart).

– Michael Servetus (Spain and Portugal, 1511–1568, veins have valved that force blood back to the heart).

– Realdo Colombo (Italy, 1510–1559, perfectly describes pulmonary circulation).

– Andrea Cesalpino (Italy, 1519–1603, circulation, role of the heart).

– William Harvey (England, 1578–1657, first complete description of the circulatory system).

– Marcello Malpighi (Italy, 1628–1694, physician, capillaries under the microscope in 1661).

Box 10.1. *Dates and origins of the scientists who have left their mark on the history of blood circulation*

Harvey's role" (taken from "The discovery of the circulation of the blood", Richet, who cites in particular Michel Servet and André Césalpin on the subject of pulmonary circulation).

10.3. The heart, engine of movement

10.3.1. *Cardiac muscle*

The heart is a muscle with four chambers: two ventricles and two atria, each communicating with a ventricle by valves (see Figure 10.3). Mechanically, it forms a system of two pumps connected in series, each pumping into the other via the systemic network or the pulmonary network. These two pumps can operate in a wide range of flow rates, from 5 l/min at rest to 25 l/min during exercise, but it should be noted that their flow rates must be exactly the same, except for short periods. They are arranged and operate roughly in the same way: a low-pressure chamber, the atrium, which receives blood from the veins; a high-pressure chamber, the ventricle, which sends blood into the arteries. The atria are relatively compliant, and their walls are thin. The ventricles, on the other hand, have thick, muscular walls whose fiber orientation rotates 180° from the internal face (endocardium) to the external face (epicardium), an arrangement that ensures that the cardiac muscle acts in all directions of the tangent plane.

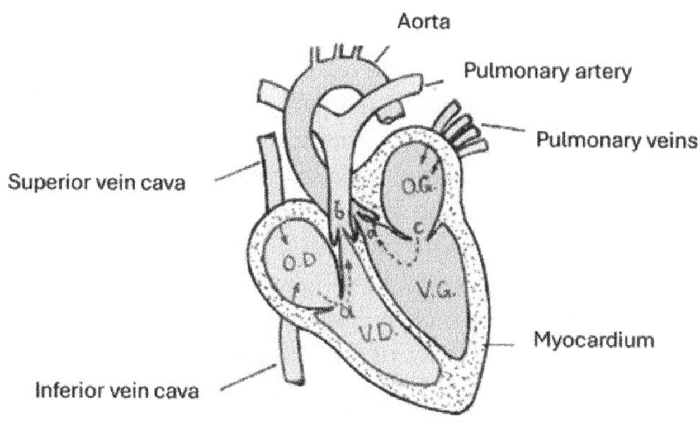

Figure 10.3. *Section of the heart, diagram according to Comolet (1984). a) Tricuspid valve; b) pulmonary valve; c) mitral valve; and d) aortic valve. For a color version of this figure, see www.iste.co.uk/prudhomme/fluid3.zip*

10.3.2. *Nodal tissue*

The contraction is caused by an excitation originating in a specialized nervous tissue (nodal tissue) of the wall of the right atrium and which propagates in the myocardial muscle at a speed of approximately 1 m/s, which means that the contractions of the different muscle fibers are relatively synchronous (Comolet 1984).

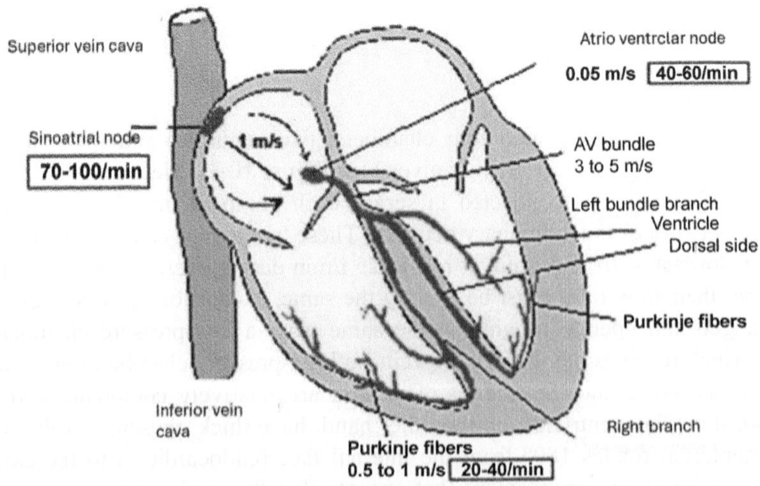

Figure 10.4. *Cardiac nodal tissue. For a color version of this figure, see www.iste.co.uk/prudhomme/fluid3.zip*

Nodal tissue (see Figure 10.4) is intraparietal cardiac tissue that gives rise to and conducts the electrical impulses that cause myocardial contraction (Mervoyer 2008/2009). It consists of the sinoatrial node known as the Keith–Flack node located in the wall of the right atrium at the level of the superior vena cava outlet. This is connected by three bundles of fibers running in the wall of the right atrium to the atrioventricular node known as the Aschoff–Tawara node at the junction of the right atrium and ventricle, close to the tricuspid septal valve and the outlet of the coronary sinus. From this node arises the trunk of the bundle of His which is subdivided into two right and left branches intended, respectively, and via the Purkinje network for the myocardial contraction of the right and left ventricles[9].

10.3.3. *Coronaries*

The heart muscle is supplied by a network of small coronary arteries from the nearby aorta[10], the return being made by a network of coronary venules (see Figure 10.5).

9 See https://en.m.wikipedia.org/wiki/File:ECG_Principle_fast.gif to follow the movements of the heart.

10 They may present narrowing's due to atheroma, spasms responsible for angina (angina pectoris) or an occlusion causing a myocardial infarction. See: https://en.wikipedia.org/wiki/Coronary_circulation.

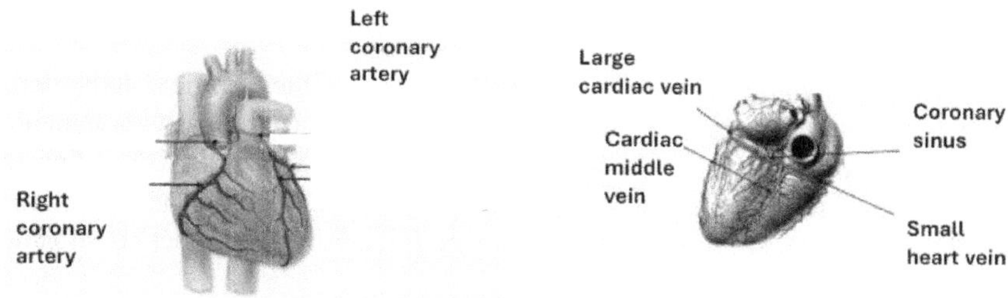

Figure 10.5. *The coronary arteries and the coronary venous network*[11]. *For a color version of this figure, see www.iste.co.uk/prudhomme/fluid3.zip*

10.3.4. *The cardiac cycle: details on diastoles and systoles (Paschal 2024)*

The cardiac cycle corresponds to a single beat and lasts about one second. It consists of four main phases:

– *Ventricular diastole*: this corresponds to the relaxation of the heart ventricles. The atria contract, which pushes blood into the ventricles. The atrioventricular valves (mitral and tricuspid valves) open, allowing blood to fill the ventricles.

– *Ventricular systole*: this is marked by the contraction of the ventricles. The atrioventricular valves close to prevent the backflow of blood into the atria. The sigmoid valves (aortic and pulmonary valves) open, allowing blood to leave the ventricles and be ejected into the aortic and pulmonary arteries.

– *Diastole*: after the ventricular systole, the ventricles relax. The sigmoid valves close to prevent the backflow of blood from the arteries into the ventricles. The atria fill again with blood from the veins.

– *Atrial systole*: just before the next cardiac cycle begins, the atria contract briefly to complete the filling of the ventricles. This ensures maximum blood flow into the ventricles before the next contraction. In short, diastole is the phase of relaxation and filling of the heart, while systole is the phase of contraction and ejection of blood. These two phases work in harmony to maintain blood flow throughout the body.

11 See: https://en.wikipedia.org/wiki/Coronary_arteries.

10.3.5. *The electrocardiogram*

An electrocardiogram is a graphical representation of the electrical activation of the heart using an electrocardiograph. This activity is collected on a lying patient, at rest, by *electrodes* placed on the surface of the skin. The standard ECG records electrical signals (*deflections*) in at least 12 *leads*, including six in the frontal plane (limb electrodes) and six in the horizontal plane (precordial electrodes) on a patient lying at rest.

Figure 10.6 shows the sequence of electrical signals recorded over time in a conventional electrocardiogram. The P, Q, R, S "waves" are observed during a cardiac cycle.

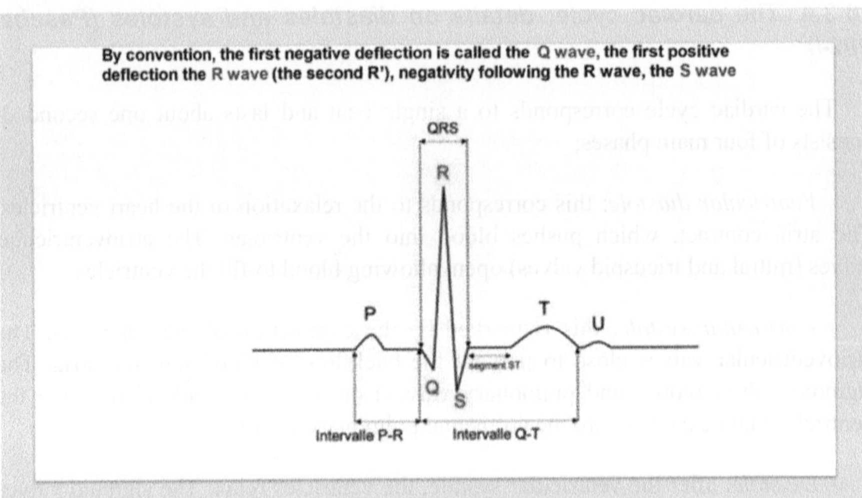

Figure 10.6. *ECG: recording of deflections during a cardiac cycle (extract from Taboulet 2025). For a color version of this figure, see www.iste.co.uk/prudhomme/fluid3.zip*

In his work, Raymond Comolet (1984, p. 71) clearly highlights the elements of the electrocardiogram with the respective pressures of the left atrium and ventricle, and of the aorta with the electrocardiogram trace. He also makes the relationships of this trace with the ventricular volumes (see Figure 10.7). Electrical activity can be printed directly onto graph paper or scanned. A standard ECG trace typically corresponds to the electrical activity of the heart for 10 seconds, or about 10 beats.

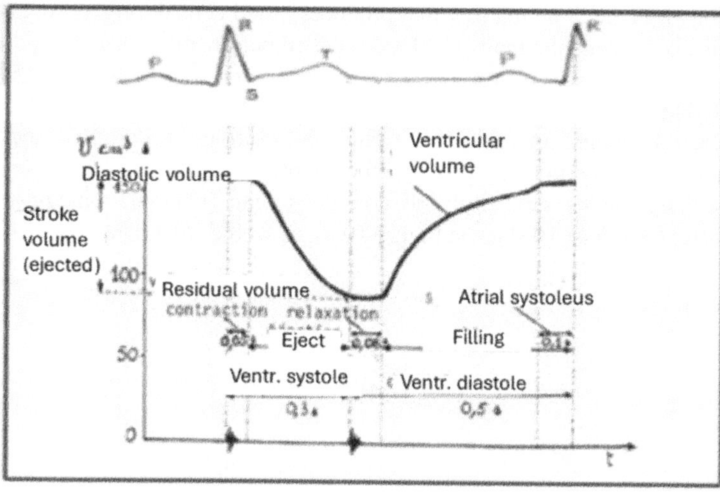

Figure 10.7. *Volumetric characteristics of the left ventricular beat (extract from Comolet 1984)*

The interest of the electrocardiogram is multiple in cardiology as underlined in Taboulet (2022), both for diagnoses (of syncopes and arrhythmias, various heart diseases) and for screening.

10.4. The fluids

10.4.1. *Blood*

Blood is a vital biological fluid that circulates continuously in the blood vessels and the heart, in particular thanks to the cardiac pump. It is composed of an aqueous fluid, plasma and a large number of cells, mainly red blood cells (erythrocytes), which give it its color and transport oxygen and carbon dioxide, as well as various white blood cells (leukocytes) participating in the body's defense and thrombocytes (platelets).

This fluid transports oxygen (O_2) and nutrients needed for vital processes in all tissues of the body, as well as wastes, such as carbon dioxide (CO_2) or nitrogenous wastes, to the sites of evacuation (kidneys, lungs, liver, intestines). It also helps transport cells and molecules of the immune system to the tissues, and diffuse hormones throughout the body.

In adults, it is the bone marrow that produces blood cells during a process called hematopoiesis. Outside the marrow, blood is called peripheral[12].

Blood rheology[13] is the subject of Chapter IV of Raymond Comolet's (1984) work, which contains, in addition to the composition of blood, a section on "Viscosimetry" and certain results on the behavioral laws that govern it: Newtonian (factors influencing its viscosity) and non-Newtonian (Fåhræus–Lindqvist effect) (Fåhræus and Lindqvist 1931) (see Figure 10.8), as well as two phase[14].

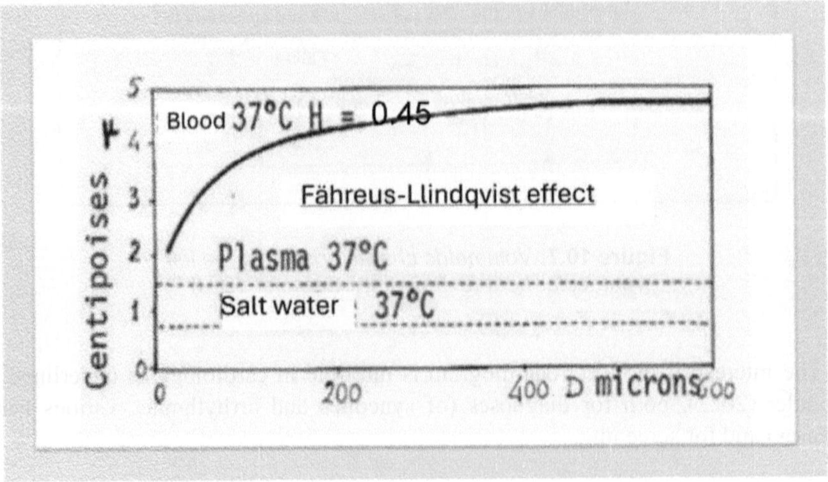

Figure 10.8. *Decrease in blood viscosity μ with a decreasing flow tube diameter D (from Comolet 1984)*

12 See: https://en.wikipedia.org/wiki/Blood.

13 Rheology is the science of the deformations and flows of matter, the resulting stresses and the efforts that must be applied to it to obtain them. It is the science of the deformations and flows of matter, the resulting stresses and the efforts that must be applied to it to obtain them. It is thus a branch of continuous media mechanics that deals with the relationships between stress, deformation and flow of matter (see: https://www.aquaportail.com/dictionnaire/definition/4681/rheologie).

14 Details on this are given in sections 10.6.5–10.7.

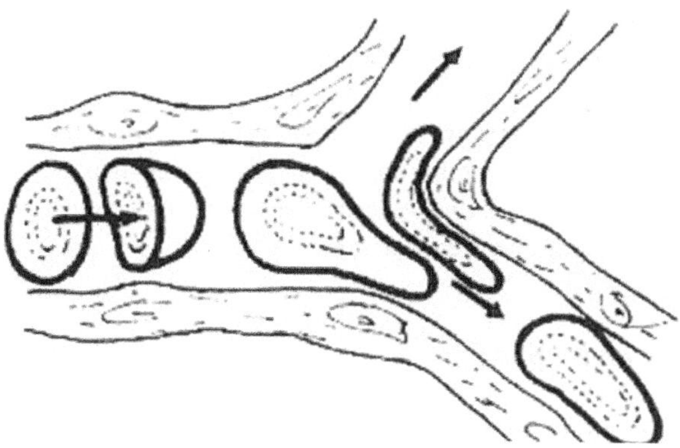

Figure 10.9. *Deformation of red blood cells penetrating the capillaries (extract from Comolet (1984))*

10.4.2. *Other fluids*

Among the fluids other than blood, we can mention the lymphatic channels[15], and among these, the cheliferous vessels.

Water, mineral salts, glucose and amino acids penetrate by diffusion into the blood capillaries of the villus. This is the blood pathway.

The products of fat digestion combine with bile salts and pass into the central chyle of the villus[16]. It is the lymphatic pathway that transports the milky-colored chyle linked to its lipid load. The chyle is released into the blood at the level of the superior vena cava. The bile salts are recovered by the liver and return to the bile.

10.5. Blood vessels

Blood circulation in the body is done by means of a multitude of vessels, different in their diameter, their parietal structures, their function. The arteries allow the expulsion of blood from the heart to the rest of the body. Oxygenated blood is projected from the left ventricle by the aorta to the muscles and organs that they irrigate by means of conduits of decreasing size, arterioles to capillaries. The blood loses its oxygen during this irrigation and is loaded with CO_2 and toxins and returns

15 See: https://en.wikipedia.org/wiki/Lymphatic_vessel.

16 Small roughness or protrusions that cover certain surfaces and give them a hairy appearance. Here, at the level of the small intestine.

to the heart (right atrium) via the capillaries, venules and veins focusing on the vena cava.

Passing from the right atrium to the right ventricle, it is then expelled by the pulmonary arteries towards the pulmonary alveoli which it penetrates via specific arterioles and capillaries. It exits towards the pulmonary veins, reaching the left atrium, then the corresponding ventricle, etc.

10.5.1. *Arteries and veins*

Figures 10.10 and 10.11 are taken from an online document from the French Federation of Cardiology[17].

An artery consists of (see Figure 10.10):

– An inner layer, the intima: composed in particular of collagen, the material that gives tissues their rigidity. It is covered with a nonstick coating: the endothelium, on which the blood and everything it carries slides.

– An intermediate layer, the media: this is the living part of the artery. Endowed with natural elasticity thanks to its smooth muscle cells, it allows the artery to modify its caliber (lumen) and its state of tension, therefore the level of blood pressure by contracting and dilating.

– An outer layer, the adventitia: this is used to anchor the arteries to the tissues they pass through.

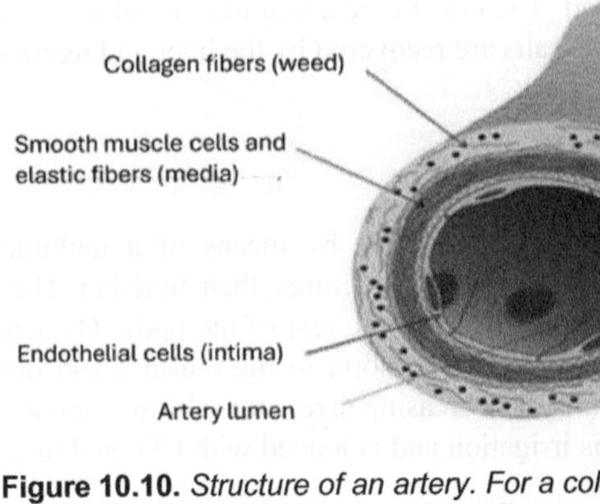

Figure 10.10. *Structure of an artery. For a color version of this figure, see www.iste.co.uk/prudhomme/fluid3.zip*

17 See: Les artères et les veines – FFC (fedecardio.org).

The veins (especially those of the lower limbs) have a considerable amount of work to do, assisted by various mechanisms (see Figure 10.11):

– Valves located along the large veins of the legs prevent blood from flowing back downwards.

– The natural action of the muscles of the legs, arms, back, stomach, etc., acts by compressing the veins when we are moving.

– When we walk, the sole of our foot, irrigated by multiple blood vessels, acts like a sponge, pushing the blood upwards with each new step.

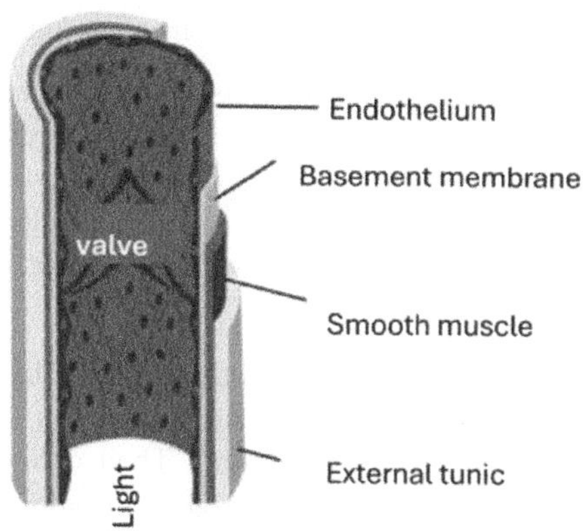

Figure 10.11. *Structure of a vein. For a color version of this figure, see www.iste.co.uk/prudhomme/fluid3.zip*

10.5.2. *Capillary vessels*

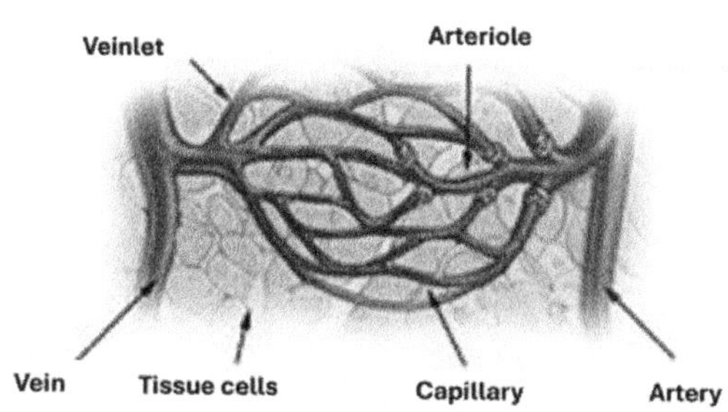

Figure 10.12. *Capillary network connecting the arterial system (right, in red) and the venous network (left, in blue). The cells are visible in orange. For a color version of this figure, see www.iste.co.uk/prudhomme/fluid3.zip*

Capillaries are the finest and smallest blood vessels found in vertebrates. They connect venules to arterioles, completing the loop of the blood circulation network (see Figure 10.12). They are arranged in tree-like networks called "capillary beds". They constitute the part of the arterial blood system where the pressure is the lowest. These networks are constantly reorganizing[18].

There are three types of blood capillaries:

1) *Continuous capillaries*: their endothelial cells are joined. They form a smooth and uninterrupted covering. They are found in skeletal muscles, smooth muscles, connective tissues, lungs (see Figure 10.13). They are sometimes enveloped by pericytes[19] whose cytoplasm has contractile proteins allowing vessel contractility.

2) *Fenestrated capillaries*: their endothelium is pierced with "micropores" (pores with a diameter of approximately 70 nm), which makes these capillaries very permeable to liquids. They are actually found where liquid transfers are necessary, in organs involved in filtration or exchanges of molecules: villi of the small intestine, renal glomeruli, endocrine glands, choroid plexuses of the cerebral ventricles, ciliary processes of the eyes, etc.

3) *Sinusoid capillaries*: their diameter is three times that of other capillaries. Their wall is uneven (and sometimes absent), nonlinear with sometimes large spaces between the endothelial cells (pores of 1–3 μm). Their cohesion is ensured by the surrounding tissues. These capillaries are therefore the most permeable to liquids and macromolecules. They are present in the liver, spleen, bones and certain endocrine glands. We also see macrophages there which phagocytose, in order to eliminate them, senescent erythrocytes or bacteria which could have crossed the skin barrier or a mucous membrane.

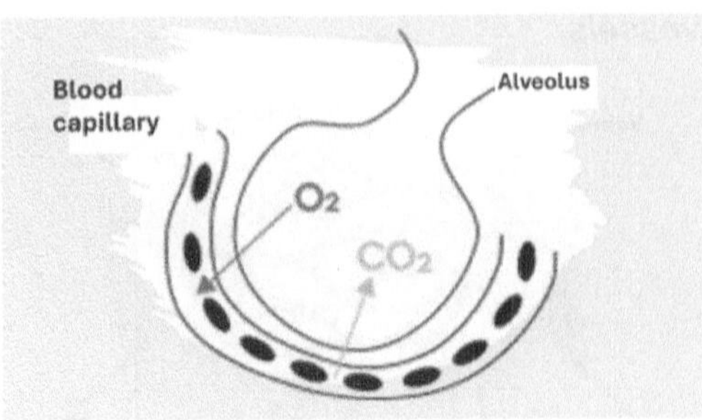

Figure 10.13. *Exchanges between the alveoli and the blood[20]. For a color version of this figure, see www.iste.co.uk/prudhomme/fluid3.zip*

18 See: https://en.wikipedia.org/wiki/Capillary.

19 Perivascular cells.

20 See: https://en.wikipedia.org/wiki/Lung.

These capillaries have a shape and structure that promote the transfer of organic macromolecules from the vascular blood to the fluid of the interstitial compartment of tissues (and vice versa). They are found in the liver, spleen, adenohypophysis, parathyroid glands, red bone marrow.

10.6. Fluid mechanics and the human body

There are many works on continuous media mechanics, in particular by Germain (1962, 1973, 1987) and more specifically of single-component fluid mechanics, forming as many tools to deal with the circulation of fluids in the human body with simplifying hypotheses. Other publications deal with homogeneous or heterogeneous multi-component flows (Prud'homme 2012, 2010).

A first question concerns whether there is any interest in completely modeling the circulation of fluids in the human body; and another is whether it is possible even partially.

As for the blood fluid, we have been able to see the complexity of the existing configurations including blood vessels and the heart.

Organs	Weight	Resting blood flow	Percentage of cardiac output
Brain	1,400	750 (1,500)	14
Heart	300	250 (1,200)	5
Liver	1,500	1300 (5,000)	23
Alimentary canal	2,500	1000 (4,000)	
Loins	300	1200 (1,500)	22
Muscles	3,500	1000 (20,000)	18
Skin	2,000	200 (4,000)	4
Skeleton, connective tissue	27,000	800 (400)	14
Total	70,000	5,500	100

Table 10.1. *Distribution of blood in the different organs of the systemic network, for a human at rest weighing 70 kg. In the third column, the values in parentheses give the blood flow in the event of maximum vasodilation*

Then, it must be seen that the flow is neither permanent nor completely pulsating and directly occupies only a part of the 15 liters of blood contained in the body with a cardiac output of approximately five and a half liters per minute (ensuring the necessary exchanges for life). Table 10.1 provided by Comolet (1984) after Folckow and Neil (1971) demonstrate the diversity of flow rates depending on the organs irrigated.

In the following, we will limit ourselves to the presentation of equations and resolutions that can be used to describe parts of the circulatory system. We will also give models developed by different authors.

10.6.1. *Basic equations*

We will not develop here the establishment of the classical equations of fluid mechanics. These are assumed to be known in volume and at the limits, as well as the balance equations of the interfaces.

The basic equations presented in this section are applicable to blood flow in vessels only at the cost of drastic approximations. Each of the assumptions adopted only concerns specific areas of the vascular system.

For example, the Bernoulli equations presented in section 10.6.2 can be applied in vessels with large diameters compared to the size of the red blood cells.

We can then consider simple fluids (i.e. fluids that can be assimilated to fluids with a single constituent). We then find very useful laws, allowing us to directly link pressure to speed.

$\vec{\mathbf{v}}$ pressure p and density ρ, subjected to the force field $\vec{f}$ at a distance per unit volume. Another simplification that may be useful is that of one-dimensional flows, presented in section 10.6.3.

In the presence of dissipations (viscosity, diffusion), the equations become more complex. Finally, when the diameters of the vessels become comparable to those of the red blood cells, the hypothesis of the continuity of the flow disappears. This is the case for capillary vessels (see Figure 10.9).

10.6.2. *Bernoulli's equations*

These equations concern fluids with a single non-viscous component, called perfect fluids, in a laminar flow, i.e. nonturbulent.

10.6.2.1. *General*

We start from the equation of the quantity of movement: $\rho\frac{d\vec{\mathbf{v}}}{dt}+\vec{\nabla}p=\rho\vec{f}$ and we assume that the remote force field derives from a potential: $\vec{f}=-\vec{\nabla}V$. We deduce from this:

$$\frac{\partial\vec{\mathbf{v}}}{\partial t}+\overrightarrow{\mathbf{rot}}\,\vec{\mathbf{v}}\times\vec{\mathbf{v}}+\vec{\nabla}\left[\frac{\vec{\mathbf{v}}^2}{2}+\int\frac{dp}{\rho}+V\right]=\vec{\mathbf{0}} \qquad [10.1]$$

Let us ask[21]:

$$\mathcal{H}=\frac{\vec{\mathbf{v}}^2}{2}+\int\frac{dp}{\rho}+V \qquad [10.2]$$

The equation for momentum becomes:

$$\frac{\partial\vec{\mathbf{v}}}{\partial t}+\overrightarrow{\mathbf{rot}}\,\vec{\mathbf{v}}\times\vec{\mathbf{v}}+\vec{\nabla}\mathcal{H}=\vec{\mathbf{0}} \qquad [10.3]$$

It is from this equation that we deduce Bernoulli's two theorems.

10.6.2.2. *Stationary case:* $\partial\vec{\mathbf{v}}/\partial t=\vec{\mathbf{0}}$.

We multiply scalarly by $\vec{\mathbf{V}}$ the two members of equation [10.3], and we then obtain:

$$\vec{\mathrm{V}}\cdot\vec{\nabla}\mathcal{H}=0\text{, or }\frac{D\mathcal{H}}{Dt}=0 \qquad [10.4]$$

The quantity $\mathcal{H}$ is constant along the streamlines. This is true whether the flow is irrotational or not, but it is stationary. This is Bernoulli's first theorem.

10.6.2.3. *Irrotational unsteady case*

This is unsteady (but the property will also be valid in stationary). In this case $\vec{\nabla}\times\vec{\mathbf{v}}=\vec{\mathbf{0}}\rightarrow\vec{\mathbf{v}}=\vec{\nabla}\phi$, and:

$$\frac{\partial\varphi}{\partial t}+\mathcal{H}=C(t) \qquad [10.5]$$

This is *Bernoulli's second theorem.*

---

21 Note that if $p=p(\rho)$ (barotropic fluid), the enthalpy is: $h=\int dp/\rho$.

10.6.2.4. *Remarks*

One of the forms of the energy equation is for perfect fluids:

$$\rho \frac{d\left(h+\vec{\mathbf{v}}^2/2\right)}{dt} = \frac{\partial p}{\partial t} + \rho \vec{\mathbf{v}} \cdot \vec{f}$$

– If the force at a distance derives from a potential $\vec{f} = -\vec{\nabla} V$ and the flow is stationary, then: $H = \frac{v^2}{2} + \int \frac{dp}{\rho} + V = Const$ on the streamlines. Since there is isoentropy (perfect fluid): $dh = T\,ds + \frac{1}{\rho}dp = \frac{1}{\rho}dp \Rightarrow h = \int \frac{dp}{\rho}$ (same as before). There is therefore identity between the stationary Bernoulli equation and the equation of conservation of energy in enthalpy form. However, if the fluid is incompressible, we do not write the equation of energy and we take $\mathcal{H} = \frac{\vec{\mathbf{v}}^2}{2} + \frac{p}{\rho} + V$.

– Unsteady: $\frac{d\left(h+\vec{\mathbf{v}}^2/2\right)}{dt} + \vec{\mathbf{v}} \cdot \vec{\nabla} V - \frac{1}{\rho}\frac{\partial p}{\partial t} = 0$

10.6.3. *The one-dimensional approximation*

In the hypothesis of a one-dimensional flow (also called quasi-one-dimensional), the flow variables are almost uniform in any straight section. These variables are therefore functions of x and t, respectively, abscissa curvilinear and time.

In the continuous state, the basic equations are as follows for an incompressible fluid without dissipative phenomena:

– constant density $\rho = Const$

– continuity $\rho\ U\ A = \dot{m} = Const$ with $U = U(x,t)$, $A = A(x,t)$

– energy: $h + \frac{U^2}{2} = h_0 \rightarrow \frac{1}{\rho} + U\,dU = 0$

This amounts to considering a current tube of variable straight section in space to take into account the spatial evolution along the blood vessel, as well as its temporal evolution, which is the case, for example, in the aorta at the exit of the heart, which dilates and retracts during a ventricular pulsation:

– Consequences for the blood: the blood in the arteries and veins can be considered as an incompressible fluid ($\rho = Const$).

10.6.4. *Viscous fluid problems*

10.6.4.1. *Dissipative simple fluids*

The pressure tensor of a fluid is written (Ventre et al 2020, section 1.4.8) as:

$$\vec{\vec{\mathbf{P}}} = p\vec{\vec{\mathbf{1}}} + (\pi - p)\vec{\vec{\mathbf{1}}} + \left(\vec{\vec{\mathbf{P}}} - \pi\vec{\vec{\mathbf{1}}}\right),\ \pi = 1/3\,\mathbf{tr}\left(\vec{\vec{\mathbf{P}}}\right) \qquad [10.6]$$

where p: thermodynamic pressure; $(\pi - p)$: volume viscosity pressure; $\vec{\vec{\mathbf{P}}} - \pi\vec{\vec{\mathbf{1}}}$: dynamic viscosity pressure tensor.

We write, for a Newtonian fluid, $\vec{\vec{\mathbf{D}}} = \left(\vec{\nabla} \otimes \vec{\mathbf{v}} + \vec{\nabla} \otimes \vec{\mathbf{v}}\right)/2$ being the tensor of the strain rates:

$$\pi - p = -k\vec{\nabla} \cdot \vec{\mathbf{v}},\ \vec{\vec{\mathbf{P}}} - \pi\vec{\vec{\mathbf{1}}} = -2\mu\left(\vec{\vec{\mathbf{D}}} - 1/3\vec{\nabla} \cdot \vec{\mathbf{v}}\vec{\vec{\mathbf{1}}}\right) \qquad [10.7]$$

Fourier's law for heat exchanges is also accepted:

$$\vec{\mathbf{q}} = -\lambda\vec{\nabla}T \qquad [10.8]$$

These relations can be deduced directly from the linear ITP[22], according to which the generalized flows (here $\pi - p$, and $\vec{\mathbf{q}}$) involved in the expression of the entropy production rate:

$$\dot{W}_S = -\frac{\pi - p}{T}\vec{\nabla} \cdot \vec{\mathbf{v}} + \vec{\mathbf{q}} \cdot \vec{\nabla}\left(\frac{1}{T}\right) - \frac{\left(\vec{\vec{\mathbf{P}}} - \pi\vec{\vec{\mathbf{1}}}\right)}{T} : \left(\vec{\vec{\mathbf{D}}} - 1/3\vec{\nabla} \cdot \vec{\mathbf{v}}\vec{\vec{\mathbf{1}}}\right) \qquad [10.9]$$

are linear functions of conjugated generalized forces, with a separation of tensor orders.

22 Irreversible thermodynamics of processes.

10.6.4.2. *Viscous incompressible fluid*

Let us use **U** of components *u, v, w* following *x, y, z* to denote the speed. We have: $\rho = Const$ or $\vec{\nabla}\cdot\vec{\mathbf{U}} = 0$ and it is generally accepted that T = Const. Therefore, μ = Const.

The laws of behavior are written:

$$p-\pi=-k\vec{\nabla}\cdot\vec{\mathbf{U}}=0,\ \vec{\vec{\mathbf{P}}}-\pi\vec{\vec{\mathbf{1}}}=-2\mu\vec{\vec{\mathbf{D}}} \qquad [10.10]$$

with: . $\underline{\underline{\mathbf{D}}}=\begin{Vmatrix} u_{,x} & \dfrac{u_{,y}+v_{,x}}{2} & \dfrac{u_{,z}+w_{,x}}{2} \\ \dfrac{u_{,y}+v_{,x}}{2} & v_{,y} & \dfrac{v_{,z}+w_{,y}}{2} \\ \dfrac{u_{,z}+w_{,x}}{2} & \dfrac{v_{,z}+w_{,y}}{2} & w_{,z} \end{Vmatrix}$

as: $\underline{\nabla}\cdot\underline{\underline{\mathbf{D}}}=\begin{Vmatrix} u_{,xx}+\dfrac{1}{2}\left(u_{,yy}+v_{,xy}\right)+\dfrac{1}{2}\left(u_{,zz}+w_{,xz}\right) \\ \dots \\ \dots \end{Vmatrix}$, and:

$$\begin{cases} u_{,x}+v_{,y}+w_{,z}=0 \Rightarrow u_{,xx}+v_{,xy}+w_{,xz}=0 \\ \left(\vec{\nabla}\cdot\vec{\vec{\mathbf{D}}}\right)_x = u_{,xx}+\dfrac{1}{2}u_{,yy}+\dfrac{1}{2}u_{,zz}-\dfrac{1}{2}u_{,xx}=\dfrac{1}{2}\Delta u \end{cases},$$

it follows: $\underline{\nabla}\cdot\underline{\underline{\mathbf{D}}}=\dfrac{1}{2}\Delta\underline{\mathbf{U}}=\dfrac{1}{2}\begin{Vmatrix}\Delta u\\ \Delta v\\ \Delta w\end{Vmatrix}$ and $\vec{\vec{\mathbf{P}}}-\pi\vec{\vec{\mathbf{1}}}=-\mu\Delta\vec{\mathbf{U}}$; hence, the Newtonian equations of viscous incompressible fluids, obeying Fourier's law:

$$\begin{aligned} &\vec{\nabla}\cdot\vec{\mathbf{U}}=0 \\ &\rho\frac{d\vec{\mathbf{U}}}{dt}+\vec{\nabla}p=\mu\Delta\vec{\mathbf{U}}+\rho\vec{f} \\ &\rho\frac{de}{dt}-\lambda\Delta T=r-\vec{\nabla}\otimes\vec{\mathbf{U}}:\vec{\vec{\mathbf{P}}} \end{aligned} \qquad [10.11]$$

Generally, in the absence of external heat input, we are not interested in the energy equation. If we also assume $\vec{f} = \vec{0}$, we obtain the incompressible Navier–Stokes equations:

$$\begin{cases} \vec{\nabla} \cdot \vec{\mathbf{U}} = 0 \\ \rho \dfrac{d\vec{\mathbf{U}}}{dt} + \vec{\nabla} p = \mu \Delta \vec{\mathbf{U}} \end{cases} \quad [10.12]$$

10.6.4.3. *Poiseuille flow*

It is the stationary flow of a fluid in a rigid cylindrical pipe (see Figure 10.14).

We consider cylindrical symmetry and a stationary flow with parallel streamlines. The problem is dealt with in cylindrical coordinates (the notation $\vec{\mathbf{U}}$ is used for the velocity). The streamlines of the flow, which is stationary, are straight parallel lines. Pressure drop Δp is the pressure drop between the inlet and outlet of the length pipe Δl. It is proposed to determine the axial velocity, and the pressure drop coefficient of the flow.

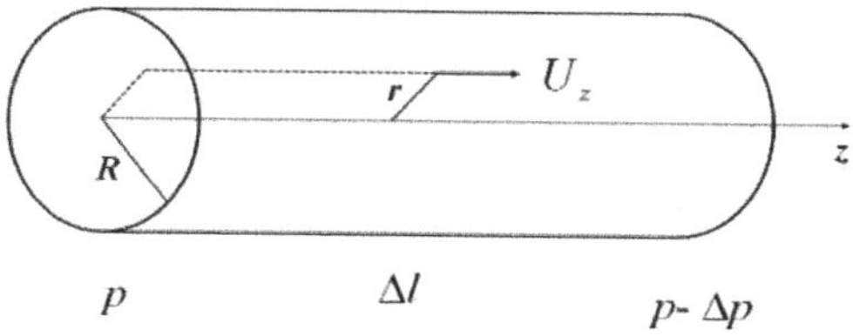

Figure 10.14. *Configuration of the Poiseuille flow*

In cylindrical coordinates: $x_1 = r, \quad x_2 = \theta, \quad x_3 = z,$ the equations are as follows.

Tensor of strain rates:

$$\begin{cases} D_{rr} = \dfrac{\partial U_r}{\partial r}, \ D_{\theta\theta} = \dfrac{1}{r}\dfrac{\partial U_\theta}{\partial \theta} + \dfrac{U_r}{r}, \ D_{zz} = \dfrac{\partial U_z}{\partial z} \\ D_{\theta z} = \dfrac{1}{2}\left(\dfrac{1}{r}\dfrac{\partial U_z}{\partial \theta} + \dfrac{\partial U_\theta}{\partial z}\right), \ D_{rz} = \dfrac{1}{2}\left(\dfrac{\partial U_r}{\partial z} + \dfrac{\partial U_z}{\partial r}\right) \\ D_{r\theta} = \left(\dfrac{\partial U_\theta}{\partial r} - \dfrac{U_\theta}{r} + \dfrac{1}{r}\dfrac{\partial U_r}{\partial \theta}\right) \end{cases} \quad [10.13]$$

Continuity: $\frac{\partial \rho}{\partial t}+\frac{1}{r}\frac{\partial(\rho r U_r)}{\partial r}+\frac{1}{r}\frac{\partial(\rho U_\theta)}{\partial \theta}+\frac{\partial(\rho U_z)}{\partial z}=0$

Stationary incompressible Navier–Stokes equations:

$$\begin{cases} \rho\left(\frac{dU_r}{dt}-\frac{U_\theta^{\,2}}{r}\right)+\frac{\partial p}{\partial r}=\rho f_r+\mu\left(\Delta U_r-\frac{2}{r^2}\frac{\partial U_\theta}{\partial \theta}-\frac{U_r}{r^2}\right) \\ \rho\left(\frac{dU_\theta}{dt}-\frac{U_r U_\theta}{r}\right)+\frac{1}{r}\frac{\partial p}{\partial \theta}=\rho f_\theta+\mu\left(\Delta U_\theta+\frac{2}{r^2}\frac{\partial U_r}{\partial \theta}-\frac{U_\theta}{r^2}\right) \\ \rho\frac{dU_z}{dz}+\frac{\partial p}{\partial z}=\rho f_z+\mu\Delta U_z \end{cases} \qquad [10.14]$$

The resolution gives the axial velocity of this laminar flow: $\mathrm{U_z}=\frac{1}{4\mu}\frac{\Delta p}{\Delta l}\left(R^2-r^2\right)$, and the pressure drop coefficient: $\Lambda=4R\frac{\Delta p}{\Delta l}\frac{1}{\rho\bar{U}^2}=\frac{32\mu}{\rho\bar{U}R}=\frac{64}{\mathrm{Re}}$, where *Re* is the Reynolds number built on the mean velocity.

10.6.4.4. *Periodic flow in a rigid pipe*

One of the simplest unsteady problems is that of periodic flow in a circular cylindrical rigid pipe. Comolet (1984) reports the most significant results provided by Womersley (1955) and Uchida (1956).

The blood is then subjected to harmonic fluctuations in shape along a rectilinear artery.

10.6.5. ***Multiphase fluids***

Blood, sometimes considered as a simple fluid, is in fact a composite fluid with several phases (see section 10.2.3.1) and deserves to be treated as such.

We now have systems of equations and methods for solving multiphase flows. At a minimum, blood could be considered a two-phase fluid if plasma and red blood cells are taken into account, which are its main constituents of obvious mechanical interest. On the pathological level, we should also be interested in the dynamics of platelets, which are likely to aggregate and thus prepare the formation of clots.

We will indicate here only the approach used to obtain the balance equations of particle suspensions (solid or liquid) in a gaseous phase (Prud'homme 2014), which can be used by adapting it to the case of interest.

First, we establish the equations governing the individual particles in motion and diffusive exchanges with the surrounding fluid. From this the balances of these particles in the volume of the mixture are deduced. We then write the equations for the conservation of the mixture (particles + fluid). By subtracting the equations of the particle system from those of the mixture, we find the equations governing the fluid. By taking into account the laws of behavior of each phase and the exchanges between phases, we obtain the coherent system to be solved given the boundary conditions.

10.6.6. *Recent developments, viscoelastic fluids*

Blood, as we have seen, is a complex multiphase fluid. It flows into vessels of different sizes with complicated walls, more or less deformable and sometimes semi-permeable.

Nevertheless, it is common to make a monophasic approximation to describe our movements. In addition, we have considered in sections 10.6.1 and 10.6.2 perfect fluids and in section 10.4.6 a viscous fluid.

More recently, other laws of behavior have been used, especially in capillary vessels. This is mainly viscoelastic rheology[23]. These laws make it possible to address both fluid flow and the formation of clots, as Ghiringhelli (2014) does, for example. The theory of viscoelastic fluids has been the subject of many books and articles. Recent articles concern applications to blood in vessels (Ventre et al. 2021).

Julien et al. (2020) study a retinal microvascular network. The hemodynamic variables of each segment obey the 1D equations deduced from the integration of the axisymmetric Navier–Stokes equations. They are coupled with a viscoelastic wall model and expressed as a function of the area of the straight section A and the flow rate Q of each micro vessel. The hemodynamic variables of each segment obey the 1D equations deduced from the integration of the axisymmetric Navier–Stokes equations. They are coupled with a viscoelastic wall model and expressed as a

23 *Viscoelasticity* is the property of materials that have both viscous and elastic characteristics when they undergo deformation. In rheology, the behavior of a linear *viscoelastic* material is thus intermediate between that of an ideal *elastic solid* symbolized by a spring of modulus E (or G) and that of a Newtonian viscous liquid symbolized by a viscosity damper η. See: https://en.wikipedia.org/wiki/Viscoelasticity.

function of the area of the straight section A and the flow rate Q of each micro vessel according to:

$$\frac{\partial A}{\partial t}+\frac{\partial Q}{\partial x}=0, \quad \frac{\partial Q}{\partial t}+\frac{\partial}{\partial x}\left(\frac{Q^2}{A}+\frac{\beta}{3\rho}A^{\frac{3}{2}}\right)=-C_f\frac{Q}{A}+C_v\frac{\partial^2 Q}{\partial x^2} \qquad [10.15]$$

where $A=2\pi\int_0^R r\,dr$, $Q=2\pi\int_0^R u_x r\,dr$, ρ is the density of the fluid and C_f the coefficient of friction. Coefficients $\beta=\frac{\sqrt{\pi}Eh}{(1-\nu^2)A_0}$ and $C_v=\frac{\sqrt{\pi}h\phi}{2\rho(1-\nu^2)\sqrt{A_0}}$ contain the physical parameters of the problem. E is Young's modulus, ν is the Poisson ratio [24], ϕ is the viscoelastic coefficient, A_0 is the reference area of straight section, and h is the arterial thickness. The authors also take into account the variation of viscosity in the microcirculation. Ventre et al. (2022) compare simplified models for calculating pressure drop through stenosis (Ghigo et al. 2018). In particular, they propose a 1D model of unsteady blood flow in the arteries, including viscoelasticity and blood cell aggregation/disintegration kinetics.

There are also articles on the behavior of vesicles simulating red blood cells for studies under more or less reduced gravity (Chaouqi 2012; Zhang et al. 2018; Danker et al. 2008).

10.7. Resolutions

Above, we have presented systems of basic equations that can be used to solve blood circulation problems as well as approximations such as one-dimensional description. A stationary viscous flow resolution was made in a rigid conduit giving the Poiseuille flow.

But blood circulation is periodic, and blood vessels are deformable. Biomechanics such as Comolet (1984) therefore first tackled periodic flows in rigid conduits, then deformable pipes.

Many problems arise for cardiologists (see, for example, François Boustani's book aimed at specialist practitioners (Boustani 2022), but which may also be of interest to biomechanics) who have to treat pathologies such as arthrosclerosis, atherosclerotic plaques, cardiac arrhythmias, aneurysm ruptures.

24 In continuum mechanics, the Poisson ratio or principal Poisson coefficient, denoted, is a dimensionless elastic constant measuring the deformation of a solid material perpendicular to the direction of the applied force.

On the other hand, astronauts are confronted with pathologies resulting from long-term stays in microgravity. Fighter pilots and aerial acrobats are subjected to very strong accelerations that can affect their blood circulation and affect the good condition of their brains. There is therefore a whole field to explore concerning the effects of acceleration and microgravity in aircraft, during weightless flights, space missions.

10.8. Conclusion

This chapter has covered the basics of cardiology, focusing on blood circulation in the human body. The fluid mechanic may have found his interest in the effort to model blood flow that is mentioned there. Multiple interfaces have been highlighted: walls of the heart, arteries, veins, capillary vessels, with their multiple exchanges at the level of the lungs and muscles and organs.

Reading this chapter may leave a justified feeling of incompleteness. This is perhaps an opportunity to suggest the writing of a fluid mechanics work on the subject, starting, for example, with Comolet's 1984 book, by making it benefit from some recent contributions from the mechanical and cardiological sciences.

Finally, it should be noted that there is a research group (GDR) Mecabio Santé[25].

10.9. Acknowledgements

The authors express their gratitude to those who agreed to review their proposed articles, which has improved their proposals.

10.10. References

Boustani, F. (2007). *La Circulation du sang : entre Orient et Occident : l'histoire d'une découverte*. Éditions Philippe Rey, Paris.

Boustani, F. (2022). *L'essentiel en cardiologie*, 3rd edition. Éditions Sauramps Medical, Montpellier [Online]. Available at: www.livres-medicaux.com.

Chaouqi, M. (2012). Vesicles, capsules and red blood cells under flow. *J. Phys.: Conf. Ser.*, 392, 012005. doi: 10.1088/1742-6596/392/1/012005.

Comolet, R. (1984). *Biomécanique circulatoire*. Masson, Paris.

Danker, G., Vlahovska, P.M., Misbah, C. (2008). Vesicles in a Poiseuille flow. *Phys. Rev. Lett.*, 102, 148102. doi: 10.1103/PhysRevLett.102.148102.

25 See: https://www.insis.cnrs.fr/fr/les-groupements-de-recherche.

Fåhræus, R. and Lindqvist, T. (1931). The viscosity of the blood in narrow capillary tubes. *The American Journal of Physiology*, 96(3), 562–568.

Folckow, B. and Neil, E. (1971). *Circulation*. Oxford University Press, New York.

Germain, P. (1962). *Mécanique des milieux continus*. Masson, Paris.

Germain, P. (1973). *Mécanique des milieux continus*, volume 1. Masson, Paris.

Germain, P. (1987). *Mécanique*, volumes 1 and 2. Ellipses, Palaiseau.

Ghigo, A.R., Lagrée, P.-Y., Fullana, J.-M. (2018). A time-dependent non-Newtonian extension of a 1D blood flow model. *J. Non-Newton. Fluid Mech.*, 253, 36–49.

Ghiringhelli, E. (2014). Viscoélasticité du sang et du caillot. PhD Thesis, Université de Grenoble, Grenoble.

Julien, L., Ventre, J., Lagrée, P.-Y., Ramkhelawon, A., Rossant, F., Atlan, M., Paques, M., Fullana, J.-M. (2020). Retinal micro-vascular network: data and model. *Computer Methods in Biomechanics and Biomedical Engineering*, 23(S1), S151–S152. doi: 10.1080/10255842.2020.1812854.

Mervoyer, E. (2008/2009). Bases de la cardiologie. Document, UNESS [Online]. Available at: https://archives.uness.fr/sites/campus-unf3s-2014/semiologie-cardiologique/enseignement/cardiologie/site/html/1_3.html.

Paschal, C. (2024). Diastole et systole : tout savoir sur le cycle cardiaque. Passeport santé, 6 November [Online]. Available at: https://www.passeportsante.net/fr/Maux/Symptomes/Fiche.aspx?doc=diastole-systole-savoir-sur-cycle-cardiaque.

Prud'homme, R. (2010). *Flows of Reactive Fluids*. Birkhäuser, Boston. doi: 10.1007/978-0-8176-4659-2.

Prud'homme, R. (2012). *Flows and Chemical Reactions*. ISTE Ltd, London and John Wiley & Sons, New York.

Prud'homme, R. (2014). *Flows and Chemical Reactions in Heterogeneous Mixtures*. ISTE Ltd, London and John Wiley & Sons, New York.

Prud'homme, R., Hounkonnou, M.N., Legros, G. (2023). L'air à feu : une histoire de flamme. *Thermodynamique des interfaces et mécanique des fluides*, 6(1). doi: 10.21494/ISTE.OP.2023.1053.

Taboulet, P. (2025). ECG. *e-cardiogram*, 25 March [Online]. Available at: https://www.e-cardiogram.com/ecg-generalites/.

Ventre, J., Abou Taam, S., Fullana, J.-M., Lagrée, P.-Y. (2021). Distribution of flow in an arteriovenous fistula using reduced-order models. *J. Biomech. Eng.*, 143(10), 101010. doi: 10.1115/1.4051282.

Ventre, J., Fullana, J.-M., Lagrée, P.-Y., Raimondi, F., Boddaert, N. (2022). Reduced-order models of blood flow: Application to arterial stenoses. In *Biological Flow in Large Vessels: Dialog Between Numerical Modeling and In Vitro/In Vivo Experiments*, Deplano, V., Fullana, J.-M., Verdier, C. (eds). ISTE Ltd, London and John Wiley & Sons, New York.

Womersley, J.R. (1955). Method for the calculation of velocity, rate of flow and viscous drag in arteries when the pressure gradient is known. *J. Physiol.*, 127(3), 553–563.

Zhang, H., Shen, Z., Hogan, B., Barakat, A.I., Misbah, C. (2018). ATP release by red blood cells under flow: Model and simulations. *Biophys. J.*, 115(11), 2218–2229.

11

Series Solutions to Boundary Value Problems for Reaction–Diffusion Equations on Finite Intervals: Application to Stefan Problems

This chapter addresses Stefan problems with two moving edges by means of a generalized series solution of the reaction–diffusion equation. The equation is first considered in its general form on an arbitrary bounded interval, and subject in a unified way to the three classical boundary conditions, namely the Dirichlet, Neumann and Robin conditions. By applying the Fourier decomposition method and Duhamel's principle, a series solution can be obtained where two main cases can be distinguished. The method is implemented on a two-sided contracting Stefan problem that describes the dynamic interaction of polymers with (or without) a substrate in the dynamic pinning model. The results show good agreement with some theoretical results. However, a particular situation involving variation in density emerges that may necessitate further theoretical investigations.

11.1. Introduction

Initial-boundary value problems for linear parabolic equations are widely used in mathematical modeling and engineering disciplines. They are used as first approximations to time-dependent systems evolving in moving domains as for Stefan problems, which are considered highly nonlinear. Other common applications are heat conductivity and diffusivity in a body (see Kumari et al. 2020), heat diffusion in multilayers with time-varying boundary and general interface conditions as (Carr and March 2018; Hickson et al. 2009), modeling calcium ions and proteins in a moving dendritic spine (Joshi and Jha 2021), oxygen and nutrient diffusion,

Chapter written by Edoh TOSSOU, Kwassi ANANI and Roger PRUD'HOMME.

metabolism dynamics and architecture optimization in a three-dimensional tissue as in McMurtrey (2016). Very often, this equation is considered over type intervals $[0,l]$ and the series solutions obtained are easily applicable to one-sided moving boundary value problems by appropriate numerical methods (see Liu et al. 2020). However, in the case of two-sided moving boundary value problems, where both edges vary simultaneously (Chayes et al. 2012; Hussein and Adil 2020), it is more appropriate to consider the problem on some generic real interval $[l_1,l_2]$ with $l_1 < l_2$. In this study, we analyze the general reaction–diffusion equation on a generic interval $[l_1,l_2]$, with the three classical boundary conditions specified in a unified way. Section 11.2 presents the general form of the problem, with brief comments on the assumptions. In section 11.3, the series solution is obtained using the Fourier decomposition method. Applications of the results are discussed in section 11.4. Finally, the conclusion is presented in section 11.5.

11.2. The reaction–diffusion equation

The most general form of a linear parabolic equation with constant coefficients can be rewritten without a convection term with an appropriate change in variables (see Henner et al. 2019). Consequently, the problem under consideration can be formulated as follows:

$$u_t(x,t) - a^2 u_{xx}(x,t) + bu(x,t) = f(x,t), \; l_1 < x < l_2, \;\; 0 < t \leq t_f \qquad [11.1]$$

$$u(x,0) = \phi(x) \qquad [11.2]$$

$$\alpha_1 u(l_1,t) + \beta_1 u_x(l_1,t) = g_1(t), \;\; \alpha_1^2 + \beta_1^2 \neq 0, \qquad [11.3]$$

$$\alpha_2 u(l_2,t) + \beta_2 u_x(l_2,t) = g_2(t), \;\; \alpha_2^2 + \beta_2^2 \neq 0, \qquad [11.4]$$

for $0 \leq t \leq t_f$. It is assumed that all intervals are open at t_f in the case where $t_f = +\infty$. The terms in equation [11.1] represent the transient, diffusion, reaction and source terms, respectively. The constant coefficient b is either a disintegration term ($b < 0$) or a multiplication term in the opposite case ($b > 0$). The coefficient $a > 0$ is the diffusivity of the mass or heat transfer. Equation [11.1] is either homogeneous if the source term $f(x, t)$ is equal to zero, or homogeneous if $f(x, t)$ is different from zero. Likewise, the initial condition [11.2] and the boundary conditions [11.3] and [11.4] can be homogeneous or inhomogeneous. The boundary conditions are expressed in a uniform way.

That is, depending on the values of the coefficients α_1, β_1, α_2 and β_2, these conditions become Dirichlet, Neumann or Robin boundary conditions. In most cases, physical conditions lead to the restriction $\alpha_1 / \beta_1 < 0$ and $\alpha_2 / \beta_2 > 0$ as discussed in Henner et al. (2019). The existence and uniqueness of the solution of Henner et al. (2009) and Marin and Ochsner 2018) follow directly from the basic principles of analysis. Before searching for a series solution, it is admitted that the source term $f(x,t)$ is continuous on $(l_1,l_2)\times(0,t_f]$, $\phi(x)$ is continuous on $[l_1,l_2]$, and $g_1(t)$, $g_2(t)$ are continuously differentiable on $[l_1,l_2]$.

Classically, the problem [11.1]–[11.4] is first solved for homogeneous equations and boundary conditions $\left(f(x,t)=g_1(t)=g_2(t)=0\right)$ using separation of variables and Fourier decomposition methods together with the Sturm–Liouville theory of self-adjoint operators. Furthermore, Duhamel's principle is applicable when a non-zero source term $f(x,t)$ is considered. Now, in the case of inhomogeneous boundary conditions, the problem [11.1]–[11.4] is first reduced to a problem with boundary conditions equal to zero by means of the so-called auxiliary functions (see e.g. Henner et al. 2019). Applying the above techniques to the problem [11.1]–[11.4] leads to an exact series solution that converges uniformly, as does the series obtained by differentiating twice by x and once by t. The uniqueness of the series solution is often proved by the maximum principle (see Guenther and Lee 2018).

11.3. The Fourier decomposition method

11.3.1. *Homogeneous equation and boundary conditions*

In this section, we will first look for a solution to the homogeneous equation:

$$u_t(x,t)-a^2u_{xx}(x,t)+bu(x,t)=0, \qquad [11.5]$$

with the initial condition:

$$u(x,0)=\phi(x), \qquad [11.6]$$

and homogeneous boundary conditions:

$$\alpha_1 u(l_1,t)+\beta_1 u_x(l_1,t)=0, \qquad [11.7]$$

$$\alpha_2 u(l_2,t)+\beta_2 u_x(l_2,t)=0. \qquad [11.8]$$

The method of variable separation assumes that the solution of the equation [11.5] can be obtained as the product of two functions of different variables x and t, i.e.:

$$u(x,t) = X(x)T(t). \tag{11.9}$$

Substituting this form into equation [11.5] yields:

$$\frac{T'(t)+bT(t)}{a^2T(t)} = \frac{X''(x)}{X(x)} = -\xi, \tag{11.10}$$

where ξ_n is a given constant. Then,

$$T'(t)+(\xi a^2+b)T(t)=0, \tag{11.11}$$

and

$$X''(x)+\xi X(x)=0. \tag{11.12}$$

The boundary conditions [11.7] and [11.8] imply for the function $X(x)$:

$$\begin{cases} \beta_1 X'(l_1)+\alpha_1 X(l_1)=0 \\ \beta_2 X'(l_2)+\alpha_2 X(l_2)=0. \end{cases} \tag{11.13}$$

According to the Sturm–Liouville theory of regular self-adjoint operators (see Guenther and Lee 2018), the solutions of the system [11.12] and [11.13] are formed by a discrete spectrum of an infinite set of nonnegative reals $\{\xi_n\}$, called eigenvalues, and the corresponding set of eigenfunctions $\{X_n(x)\}$.

Therefore, the solutions of equation [11.12] are of the form: $X_n(x) = C_1^n \cos\sqrt{\xi_n}x + C_2^n \sin\sqrt{\xi_n}x$, where C_1^n and C_2^n are unknown constant coefficients. Now, consider the general form of the solution:

$$X(x) = C_1 \cos\sqrt{\xi}x + C_2 \sin\sqrt{\xi}x, \tag{11.14}$$

where C_1 and C_2 are the unknown coefficients. Applying this to the conditions [11.13], we obtain the following system of equations in the coefficients C_1 and C_2:

$$\begin{cases} C_1\left[\alpha_1 \cos\sqrt{\xi}l_1 - \beta_1\sqrt{\xi}\sin\sqrt{\xi}l_1\right] + C_2\left[\beta_1\sqrt{\xi}\cos\sqrt{\xi}l_1 + \alpha_1\sin\sqrt{\xi}l_1\right] = 0 \\ C_1\left[\alpha_2 \cos\sqrt{\xi}l_2 - \beta_2\sqrt{\xi}\sin\sqrt{\xi}l_2\right] + C_2\left[\beta_2\sqrt{\xi}\cos\sqrt{\xi}l_2 + \alpha_2\sin\sqrt{\xi}l_2\right] = 0 \end{cases} \quad [11.15]$$

Note that this linear homogeneous system for C_1 and C_2 admits non-trivial solutions if and only if its determinant is equal to zero. In this case, the calculations yield:

$$\tan\mu = \frac{(\beta_1\alpha_2 - \alpha_1\beta_2)(l_2 - l_1)\mu}{\mu^2\beta_1\beta_2 + (l_2 - l_1)^2\alpha_1\alpha_2} \quad [11.16]$$

where $\mu = \sqrt{\xi}(l_2 - l_1)$ and $\xi = (\frac{\mu}{l_2 - l_1})^2$.

Equation [11.16] admits an infinite number of solutions μ_n and therefore generates an infinite number of real eigenvalues ξ_n. The eigenvalues are computed as:

$$\xi_n = (\frac{\mu_n}{l_2 - l_1})^2. \quad [11.17]$$

Using trigonometric transformations, the first equation of the system [11.15] takes the form:

$$\sqrt{\beta_1^2\xi_n + \alpha_1^2}\cos(\sqrt{\xi_n}l_1 + \theta_n)C_1 + \sqrt{\beta_1^2\xi_n + \alpha_1^2}\sin(\sqrt{\xi_n}l_1 + \theta_n)C_2 = 0, \quad [11.18]$$

where $\tan^{-1}(\frac{\beta_1\sqrt{\xi_n}}{\alpha_1}) = \theta_n$ and θ_n is unique when taken in the interval $\left[-\frac{\pi}{2}, \frac{\pi}{2}\right]$.

Let $A = \sqrt{\beta_1^2\xi_n + \alpha_1}\cos(\sqrt{\xi_n}l_1 + \theta_n)$ and $B = \sqrt{\beta_1^2\xi_n + \alpha_1}\sin(\sqrt{\xi_n}l_1 + \theta_n)$, equation [11.18] becomes:

$$AC_1 + BC_2 = 0 \Leftrightarrow \frac{C_1}{B} = -\frac{C_2}{A} = C, \quad [11.19]$$

with C being constant.

Here, the constants C_1 and C_2 can be deduced as:

$$C_1 = BC,\ C_2 = -AC_1 \quad [11.20]$$

Substituting the constants C_1 and C_2into equation [11.14], we obtain the eigenfunctions X_n corresponding to the eigenvalues ξ_n, as:

$$X_n(x) = C\left[B\beta_1\sqrt{\xi_n}\cos\sqrt{\xi_n}x - A\alpha_1\sin\sqrt{\xi_n}x\right]. \tag{11.21}$$

Considering now the ordinary first degree equation [11.11] for $\xi_n = \xi$, the solution is written as $T_n(t) = C_n e^{-(a^2\xi_n+b)t}$. The family of solutions satisfying equation [11.5] and boundary conditions [11.7] and [11.8] can therefore be written as:

$$u_n(x,t) = X_n(x)T_n(t) = C_n e^{-(a^2\xi_n+b)t}X_n(x). \tag{11.22}$$

According to the principle of superposition, the sum:

$$u(x,t) = \sum_{n=1}^{\infty} w_n(x,t) = \sum_{n=1}^{\infty} C_n e^{-(a^2\xi_n+b)t}X_n(x) \tag{11.23}$$

is the solution of the system equations [11.5]–[11.7] and [11.8]. The coefficients C_n are obtained using the initial condition [11.6]. In fact, according to Sturm–Liouville theory, the eigenfunctions $\{X_k(x), k \geq 1\}$ form a complete orthogonal set on the interval $[l_1, l_2]$ and the series expansion of the function $\phi(x)$ takes the form:

$$u(x,0) = \varphi(x) = \sum_{k=1}^{\infty} C_k X_k(x). \tag{11.24}$$

Due to the orthogonality condition, multiplying the relation [11.24] by $X_n(x)$ for a given $n \geq 1$ and integrating from l_1 to l_2, the coefficient C_n is given by the equation:

$$C_n = \frac{1}{\| X_n \|^2}\int_{l_1}^{l_2} \varphi(x)X_n(x)dx, \tag{11.25}$$

where the $\| X_n \|^2 = \int_{l_1}^{l_2} X_n(x)^2\,dx$, in such a way that each element $X_n(x)$ of the orthogonal basis is normalized to unity. Combining relations [11.24] and [11.25], and changing the order of summation and integration by using uniform convergence, the solution $u(x,t)$ is transformed into:

$$u(x,t)=\int_{l_l}^{l_2} g(x,\eta,t)\phi(\eta)d\eta \qquad [11.26]$$

where

$$g(x,\eta,t)=\frac{\sum_{n=1}^{\infty} X_n(x)X_n(\eta)}{\| X_n \|^2} e^{-(a^2\xi_n+b)t} \qquad [11.27]$$

11.3.2. *Non-homogeneous problem and auxiliary functions*

In this section, the reaction–diffusion equation with a local source term and inhomogeneous boundary conditions is considered. The equation has the form:

$$u_t(x,t)-a^2u_{xx}(x,t)+bu(x,t)=f(x,t),\ l_1<x<l_2,\ 0<t\le t_f \qquad [11.28]$$

Keeping the initial condition [11.6] and the homogeneous boundary conditions [11.7] and [11.8], the solution $u(x,t)$ can be decomposed into the sum of two functions:

$$u(x,t)=u_h(x,t)+u_p(x,t), \qquad [11.29]$$

where $u_h(x,t)$ satisfies the homogeneous equation [11.5] with initial condition [11.6] and homogeneous boundary conditions [11.7] and [11.8]. The function $u_p(x,t)$ satisfies the inhomogeneous equation with homogeneous boundary conditions [11.7] and [11.8] and zero initial condition, i.e.:

$$u_p(x,0)=0. \qquad [11.30]$$

As discussed in section 11.3.1, u_p can be found as:

$$u_p(x,t)=\sum_{n=1}^{\infty} X_n(x)T_n(t) \qquad [11.31]$$

We insert the formula [11.31] into equation [11.28] and note that $X_n''(x) = -\xi_n X_n(x)$. This gives:

$$\sum_{n=1}^{\infty}\left[T_n'(t) + (a^2\xi_n + b)T_n(t)\right]X_n(x) = f(x,t) \qquad [11.32]$$

According to the completeness property of Sturm–Liouville eigenfunctions (see Henner et al. 2019; Guenther and Lee 2018), the function $f(x,t)$ can be expanded into a Fourier series of functions $X_n(x)$ over the interval (l_1, l_2) as follows:

$$f(x,t) = \sum_{n=1}^{\infty} f_n(t)X_n(x). \qquad [11.33]$$

The orthogonality property of the functions X_n and the uniqueness of the decomposition [11.33] allow us to obtain the functions f_n in the form:

$$f_n(t) = \frac{1}{\| X_n \|^2}\int_{l_1}^{l_2} f(x,t)X_n(x)dx\,. \qquad [11.34]$$

Comparing formula [11.32] with [11.33], an ordinary first degree differential equation is derived as:

$$T_n'(t) + (a^2\xi_n + b)T_n(t) = f_n(t)\,, \qquad [11.35]$$

with the initial condition $T_n(0) = 0$, which comes from equation [11.30]. The solution is:

$$T_n(t) = \int_0^t f_n(\tau)e^{-(a^2\xi_n + b)(t-\tau)}d\tau\,. \qquad [11.36]$$

Writing $Y_n(t-\tau) = e^{-(a^2\xi_n + b)(t-\tau)}$, T_n becomes:

$$T_n(t) = \int_0^t f_n(\tau)Y_n(t-\tau)d\tau \qquad [11.37]$$

The solution [11.29] of the inhomogeneous equation with zero boundary conditions is now in the form:

$$u(x,t)=u_h(x,t)+u_p(x,t)=\sum_{n=1}^{\infty}\left[T_n(t)+C_n e^{-(a^2\xi_n+b)t}\right]X_n(x). \qquad [11.38]$$

Since we aim to find the series solution of the problem with inhomogeneous boundary conditions [11.1]–[11.4], it is necessary to reduce the problem to the one with zero boundary conditions as in section 11.3.1. Therefore, the solution of the problem is sought in the form:

$$u(x,t)=v(x,t)+w(x,t) \qquad [11.39]$$

where the so-called auxiliary function $w(x,t)$ is chosen in order to satisfy the inhomogeneous boundary conditions [11.3] and [11.4], i.e.:

$$\begin{cases}\beta_1 w(l_1,t)+\alpha_1 w(l_1,t)=g_1(t)\\ \beta_2 w(l_2,t)+\alpha_2 w(l_2,t)=g_2(t)\end{cases}, \qquad [11.40]$$

In the formula [11.39], $v(x,t)$ is the unknown function that satisfies the inhomogeneous equation:

$$v_t(x,t)-a^2 v_{xx}(x,t)+bv(x,t)=\tilde{f}(x,t), \qquad [11.41]$$

with the initial condition:

$$v(x,0)=\tilde{u}_0(x), \qquad [11.42]$$

with the zero boundary conditions:

$$\beta_1 v_x(l_1,t)+\alpha_1 v(l_1,t)=0, \qquad [11.43]$$

$$\beta_1 v_x(l_1,t)+\alpha_1 v(l_1,t)=0, \qquad [11.44]$$

where

$$\tilde{f}(x,t)=f(x,t)-\frac{\partial}{\partial t}w(x,t)+a^2\frac{\partial}{\partial x^2}w(x,t)-bw(x,t), \text{ and}$$

$$\tilde{u}_0(x)=u_0(x)-w(x,0).$$

The solution to the latter problem [11.41]–[11.44] has already been discussed in section 11.3.1 and we simply write:

$$v(x,t) = \sum_{n=1}^{\infty}\left[T_n(t) + C_n e^{-(a^2\xi_n + b)t}\right]X_n(x). \qquad [11.45]$$

To simplify the calculations, the auxiliary function $w(x,t)$ is constructed in the form:

$$w(x,t) = P_1(x)g_1(t) + P_2(x)g_2(t), \qquad [11.46]$$

where $P_1(x)$ and $P_2(x)$ are chosen to be the first or second degree polynomials. Two main cases can be considered, the case where the coefficients α_1 and α_2 related to the boundary conditions are not simultaneously zero (i.e. $\alpha_1 \neq 0$ or $\alpha_2 \neq 0$) and a situation where the two coefficients α_1 and α_2 are zero.

– *Case 1:* $\alpha_1 \neq 0$ or $\alpha_2 \neq 0$.

P_1 and P_2 are chosen as the first degree polynomials:

$$P_1(x) = \gamma_1 + \delta_1 x, \; P_2(x) = \gamma_2 + \delta_2 x, \qquad [11.47]$$

and the auxiliary function w takes the form:

$$w(x,t) = (\gamma_1 + \delta_1 x)g_1(t) + (\gamma_2 + \delta_2 x)g_2(t). \qquad [11.48]$$

Substituting w into the boundary conditions [11.40], and after reduction and identification, we obtain the following system of four equations with four unknown coefficients γ_1, γ_2, δ_1 and δ_2:

$$\begin{cases} \beta_1\delta_1 + \alpha_1\gamma_1 + \alpha_1\delta_1 l_1 = 1 \\ \beta_1\delta_2 + \alpha_1\gamma_2 + \alpha_1\delta_2 l_1 = 0 \\ \beta_2\delta_1 + \alpha_2\gamma_1 + \alpha_2\delta_1 l_2 = 0 \\ \beta_2\delta_2 + \alpha_2\gamma_2 + \alpha_2\delta_2 l_2 = 1 \end{cases} \qquad [11.49]$$

The solutions of the above system are:

$$\gamma_1 = \frac{\beta_2 + \alpha_2 l_2}{\beta_2\alpha_1 - \beta_1\alpha_2 + \alpha_1\alpha_2(l_2 - l_1)}, \; \gamma_2 = \frac{\beta_1 + \alpha_2 l_1}{\beta_2\alpha_1 - \beta_1\alpha_2 + \alpha_1\alpha_2(l_2 - l_1)} \qquad [11.50]$$

$$\delta_1 = \frac{-\alpha_2}{\beta_2\alpha_1 - \beta_1\alpha_2 + \alpha_1\alpha_2(l_2 - l_1)}, \delta_2 = \frac{\alpha_2}{\beta_2\alpha_1 - \beta_1\alpha_2 + \alpha_1\alpha_2(l_2 - l_1)}. \qquad [11.51]$$

From all of the above, the auxiliary function $w(x, t)$ takes the form:

$$\begin{aligned} w(x,t) = &\left(\frac{\beta_2 + \alpha_2 l_2}{\beta_2\alpha_1 - \beta_1\alpha_2 + \alpha_1\alpha_2(l_2 - l_1)} + \frac{-\alpha_2}{\beta_2\alpha_1 - \beta_1\alpha_2 + \alpha_1\alpha_2(l_2 - l_1)} x\right) g_1(t) \\ &+ \left(\frac{\beta_1 + \alpha_2 l_1}{\beta_2\alpha_1 - \beta_1\alpha_2 + \alpha_1\alpha_2(l_2 - l_1)} + \frac{\alpha_1}{\beta_2\alpha_1 - \beta_1\alpha_2 + \alpha_1\alpha_2(l_2 - l_1)} x\right) g_2(t). \end{aligned} \qquad [11.52]$$

The solution to the problem [11.1]–[11.4] takes the form:

$$u(x,t) = \sum_{n=1}^{\infty} \left[T_n(t) + C_n e^{-(a^2\xi_n + b)t} \right] X_n(x) + w(x,t). \qquad [11.53]$$

– *Case 2:* $\alpha_1 = \alpha_2 = 0$.

The boundary conditions [11.40] become:

$$\begin{cases} w_x(l_1, t) = g_1(t), \\ w_x(l_2, t) = g_2(t). \end{cases} \qquad [11.54]$$

The polynomial $P_1(x)$ and $P_2(x)$ must be of degree greater than 1, so they are chosen as:

$$P_1(x) = \gamma_1 + \delta_1 x + \lambda_1 x^2, \; P_1(x) = \gamma_2 + \delta_2 x + \lambda_2 x^2 \qquad [11.55]$$

Substituting the new form of $w(x,t)$ into the boundary conditions [11.40] leads to the system:

$$\begin{cases} \delta_1 + 2\lambda_1 l_1 = 1, \\ \delta_2 + 2\lambda_2 l_1 = 0, \\ \delta_1 + 2\lambda_1 l_1 = 0, \\ \delta_2 + 2\lambda_2 l_2 = 0. \end{cases} \qquad [11.56]$$

The resolution results in the following values of $\lambda_1, \lambda_2, \delta_1$ and δ_2 :

$$\lambda_1 = -\frac{1}{2(l_2 - l_1)} \quad \lambda_2 = \frac{1}{2(l_2 - l_1)} \qquad [11.57]$$

$$\delta_1 = \frac{l_2}{l_2 - l_1} \quad \delta_2 = -\frac{l_1}{l_2 - l_1}. \qquad [11.58]$$

Finally, taking $\gamma_1 = \gamma_2 = 0$, the generic form of $w(x,t)$ is given as:

$$w(x,t) = \frac{1}{2(l_2 - l_1)}\left[\left(l_2 x - x^2\right)g_1(t) + \left(-l_1 x + x^2\right)g_2(t)\right]. \qquad [11.59]$$

The solution of the problem [11.1]–[11.4] has the form:

$$u(x,t) = \sum_{n=1}^{\infty}\left[T_n(t) + C_n e^{-(a^2\xi_n + b)t}\right]X_n(x) + \frac{1}{2(l_2 - l_1)} \left[\left(l_2 x - x^2\right)g_1(t) + \left(-l_1 x + x^2\right)g_2(t)\right]. \qquad [11.60]$$

11.4. Applications

11.4.1. *Practical implementation of the method*

To apply the above series solutions to short durations, as in so-called time-step models in computational fluid dynamics, for example, the time interval $[0,T]$ is subdivided into N time steps with a regular step Δt such that Δt is sufficiently small, for example, $\Delta t \leq 10^{-2}$ (see among others Epperson 2013; Izadi and Yuzbasi 2022). The boundary locations l_1 and l_2 are assumed to be fixed during the time step Δt, but vary from time step to time step. Truncating the infinite series solution at some fixed order, the procedure for calculating analytical approximations of the solution during the first time step $t \in [0,\Delta t]$ can be repeated for the next time step. For this to be done, it is sufficient to update the function $\phi(x)$ from the previous step, by using the corresponding source terms $f(x,t)$ and $g_1(t), g_2(t)$ which are related to the boundary conditions. Thus, by using the Stefan conditions, the new values of the boundary locations l_1 and l_2 are repetitively computed until convergence is attained at the last time step. Therefore, one-dimensional computational models for two-sided Stefan problems can be efficiently handled when using the present series solutions. Examples of two-sided moving boundary problems can be seen in Chayes et al. (2012).

11.4.2. *A two-sided contracting Stefan problem*

Consider the following Stefan problem reporting in Lacoin (2014):

$$\begin{cases} \dfrac{\partial \varphi}{\partial t} - \dfrac{\partial^2 \varphi}{\partial x^2} = 0, \text{ on } \left(l(t), r(t)\right), \\ \varphi(x,0) = \varphi_0(x), \text{ on } \left(l_0, r_0\right), \\ \varphi(l(t),t) \equiv 1 \text{ on } [-1, l(t)],\ \varphi\left(r(t),t\right) \equiv -1 \text{ on } \left[r(t),1\right], \\ l'(t) = -\varphi_x(l(t),t),\ r'(t) = \varphi_x(r(t),t), \end{cases} \quad [11.61]$$

where $\varphi(x,t)$ is the temperature in the case of heat diffusion or the particle density in the case of mass transfer and $l(t)$, $r(t)$ are the moving edges, with $l_0 = l(0)$ and $r_0 = r(0)$. This issue arises from the mathematical model that describes the dynamic interaction of polymers with or without a substrate in the dynamic pinning model, as reported in Lacoin (2014). In the latter, it was shown that this problem has a unique classical solution up to a certain point in time:

$$T^* = \frac{1}{2}\int_{l_0}^{r_0} f_0(x)\mathrm{d}x, \quad [11.62]$$

where the two boundaries meet, provided that the function $f_0(x)$ representing the primitive function of the initial density $\varphi_0(x)$ is smooth in the interval (l_0, r_0), 1-Lipschitz positive, and satisfies the boundary conditions $f_0{}'(l_0) = -f_0{}'(r_0) = 1$.

The last two conditions of system [11.61], which give the boundary velocities, are the Stefan conditions. Solving such a system involves finding the function $\varphi(x,t)$ and the positions of the edges $l(t), r(t)$.

Assuming the edges are fixed during a time step $\Delta t = t_{i+1} - t_i$, the system is reduced to an initial-boundary value problem with Dirichlet boundary conditions as follows:

$$\begin{cases} \dfrac{\partial \varphi}{\partial t} - \dfrac{\partial^2 \varphi}{\partial x^2} = 0, \text{ for } l_i < x < r_i,\ t_i \le t \le t_{i+1}, \\ \varphi(x,0) = \varphi_0(x), \\ \varphi(l_i,t) = 1, \\ \varphi\left(r_i,t\right) \equiv -1, \end{cases} \quad [11.63]$$

where $r_i = r(t_i)$ and $l_i = l(t_i)$ for $t = t_i$, $i = 0,1,3,\ldots$. It can be verified that the function $f_0(x) = -2x + 3\ln(x+2)$ satisfies all the conditions mentioned above in order to have a unique classical solution. So, the derivative of $f_0(x)$ gives the function $\varphi_0(x) = -\dfrac{2x+1}{x+2}$. Taking this latter function as the initial condition, and (l_0, r_0) equal to $(-1,1)$, the solution of the problem [11.61] during each time step Δt is obtained from section 11.3.2, case 1, as follows:

$$\varphi(x,t) = (r_0 - l_0) - 2x + \sum_{n=1}^{\infty} C_n e^{-\xi_n t} \sin\left(\frac{n\pi x}{r_0 - l_0}\right), \qquad [11.64]$$

where $\xi_n = \left(\dfrac{n\pi}{r_i - l_i}\right)^2$ and the coefficients C_n are given by:

$$C_n = \frac{2}{r_i - l_i} \int_{l_i}^{r_i} \varphi_i(x) \sin\left(\frac{n\pi x}{r_i - l_i}\right) dx. \qquad [11.65]$$

The initial condition at time t_i is denoted by $\varphi_i(x)$ since that function changes at each time step by taking the value $\varphi(x, t_{i-1})$ of the previous time t_{i-1}.

By coupling this solution with the Stefan conditions:

$$l'(t) = -\varphi_x(l(t), t) = -\left.\frac{\partial \varphi(x,t)}{\partial x}\right|_{x=l(t)}, \qquad [11.66]$$

and

$$r'(t) = \varphi_x(r(t), t) = \left.\frac{\partial \varphi(x,t)}{\partial x}\right|_{x=r(t)}, \qquad [11.67]$$

the edges l_{i+1}, r_{i+1} can be deduced recursively using numerical approximations of derivatives as follows:

$$l_{i+1} = l_i - \Delta t \times \frac{\partial \varphi(l_i, t)}{\partial x}, \qquad [11.68]$$

$$r_{i+1} = r_i + \Delta t \times \frac{\partial \varphi(r_i, t)}{\partial x}. \quad [11.69]$$

Thus, by repeating this procedure as long as the remaining time of the process is greater than Δt, we obtain the evolution of the edges $l(t), r(t)$ and that of the density of the particles φ (x, t). In order to enhance the precision of the approximations, all series have been truncated to the first 20 terms in the numerical program by using, for example, Matlab software. Figure 11.1 is obtained using the dimensionless time step $\Delta t = 10^{-2}$, and shows the evolution of the edges $l(t)$ and $r(t)$ over time t. Figure 11.2(a) shows the evolution of the density at both edges $l(t)$ and $r(t)$, as well as the average density of the particles.

Figures 11.1 and 11.2(a) demonstrate a consistent convergence time at the point where the two boundaries meet. However, in Figure 11.2(b), which is an enlargement of Figure 11.2(a), it is clearly shown that the densities cross and change sign near the beginning of the process, well before the convergence time. This situation may warrant further theoretical investigation.

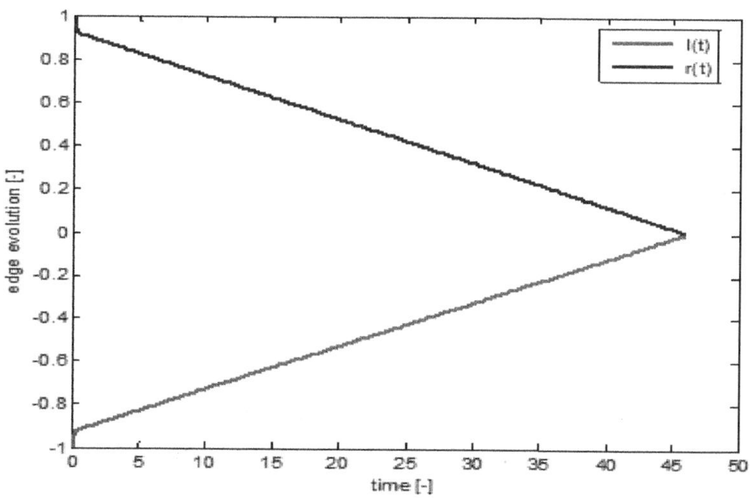

Figure 11.1. *Evolution of edges l(t) and r(t). For a color version of this figure, see www.iste.co.uk/prudhomme/fluid3.zip*

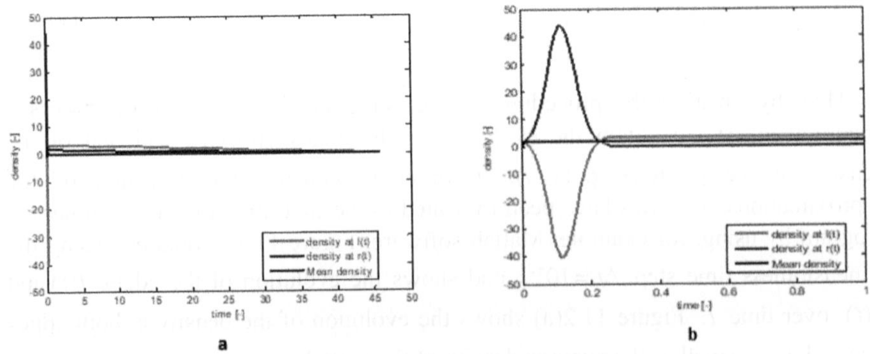

Figure 11.2. *a) Evolution of density at edges l(t) and r(t). b) Enlarge figure a on t ∈ (0,1). For a color version of this figure, see www.iste.co.uk/prudhomme/fluid3.zip*

11.5. Conclusion

This chapter has permitted the generalization of the classical series solutions for the Fourier decomposition method for reaction–diffusion equations to generic intervals on the real line. The problem is solved in its most general form, with the boundary conditions provided in a unified manner. The series solutions obtained can be used for the efficient modeling of two-sided moving edge problems. Although infinite series solutions are often inaccessible in closed form, their current generalization on arbitrary intervals of the real line may enhance numerical efficiency in one-dimensional Stefan models and also in multilayer diffusion problems.

11.6. References

Carr, E.J. and March, N.G. (2018). Semi-analytical solution of multilayer diffusion problems with time-varying boundary conditions and general interface conditions. *Appl. Math. Comput.*, 333, 286. doi: 10.1016/j.amc.2018.03.095.

Chayes, L., Kim, I., Gui, C. (2012). The supercooled stefan problem in one dimension. *Commun. Pure Appl. Anal.*, 11, 845.

Epperson, J.F. (2013). *An Introduction to Numerical Methods and Analysis*, 2nd edition. John Wiley & Sons, New York.

Guenther, R.B. and Lee, J.W. (2018). *Sturm-Liouville Problems: Theory and Numerical Implementation*, 1st edition. CRC Press, Boca Raton. doi: 10.1201/9780429437878.

Henner, V., Belozerova, T., Forinash, K. (2009). *Mathematical Models in Physics: Partial Differential Equations, Fourier Series, and Special Functions*. CRC Press, Boca Raton.

Henner, V., Belozerova, T., Nepomnyashchy, A. (2019). *Partial Differential Equations: Analytical Methods and Applications*. CRC Press, Boca Raton. doi: 10.1201/9780429440908.

Hickson, R.I., Barry, S.I., Mercer, G.N. (2009). Critical times in multilayer diffusion. Part 1: Exact solutions. *Int. J. Heat Mass Transfer*, 52, 5776. doi: 10.1016/j.ijheatmasstransfer.2009.08.013.

Hussein, M.S. and Adil, Z. (2020). Numerical solution for two-sided Stefan problem. *Iraqi J. Sci.*, 61(2), 444. doi: 10.24996/ijs.2020.61.2.24.

Izadi, M. and Yuzbasi, S. (2022). A hybrid approximation scheme for 1-D singularly perturbed parabolic convection-diffusion problems. *Math. Commun.*, 27(1), 47.

Joshi, H. and Jha, B.K. (2021). On a reaction–diffusion model for calcium dynamics in neurons with Mittag–Leffler memory. *Eur. Phys. J. Plus*, 136(6), 623. doi: 10.1140/epjp/s13360-021-01610-w.

Kumari, T., Kumar, D., Rai, K.N., Singh, S.K. (2020). Numerical solution of DPL heat transfer model in multi-layer biological skin tissue of the living body during hyperthermia treatment. *Mech. Based Des. Struct. Mach.*, 51(1), 159. doi: 10.1080/15397734.2020.1836652.

Lacoin, H. (2014). The scaling limit of polymer pinning dynamics and a one dimensional Stefan freezing problem. *Commun. Math. Phys.*, 331, 21. doi: 10.1007/s00220-014-2058-1.

Liu, S., Du, Y., Liu, X. (2020). Numerical studies of a class of reaction–diffusion equations with Stefan conditions. *Int. J. Comput. Math.*, 97(5), 959. doi: 10.1080/00207160.2019.1599868.

Marin, M. and Ochsner, A. (2018). *Essentials of Partial Differential Equations*, 1st edition. Springer, Cham. doi: 10.1007/978-3-319-90647-8.

McMurtrey, R.J. (2016). Analytic models of oxygen and nutrient diffusion, metabolism dynamics, and architecture optimization in three-dimensional tissue constructs with applications and insights in cerebral organoids. *Tissue Eng. Part C Methods*, 22(3), 221. doi: 10.1089/ten.tec.2015.03.

List of Authors

Tsialefitry ALY SAANDY
Laboratoire de l'Energie Renouvelable et Environnement
l'E.S. Polytechnique
Université d'Antsiranana
Madagascar

Kwassi ANANI
Laboratory of Mathematical Modelling and Applications
Department of Mathematics
Faculty of Sciences
University of Lomé
Togo

François BOUSTANI
Independent Cardiologist and Historian
Montrouge
France

Jean-Paul CALTAGIRONE
Bordeaux INP
Arts et Métiers Institute of Technology
INRAE, CNRS UMR 5295
I2M Bordeaux
France

Christian CHAUVEAU
ICARE, CNRS
Orléans
France

Mahouton Norbert HOUNKONNOU
International Chair in Mathematical Physics and Applications (ICMPA UNESCO Chair)
University of Abomey-Calavi
Republic of Benin

David ILUNGA
Département d'Electromécanique
Faculté Polytechnique
Université de Lubumbashi
Democratic Republic of the Congo

Jean-Paul KATON MBAY
Département d'Electromécanique
Faculté Polytechnique
Université de Lubumbashi
Democratic Republic of the Congo

Guillaume LEGROS
Sorbonne Université
Paris
France

Deo MWELWA
Département d'Electromécanique
Faculté Polytechnique
Université de Lubumbashi
Democratic Republic of the Congo

Roger PRUD'HOMME
Jean Le Rond d'Alembert Institute
UMR 7190, CNRS
Sorbonne University
Paris
France

Francis RAVELOSON
Laboratoire de l'Energie
Renouvelable et Environnement
l'E.S. Polytechnique
Université d'Antsiranana
Madagascar

Delphin TOMBORAVO
Laboratoire de l'Energie
Renouvelable et Environnement
l'E.S. Polytechnique
Université d'Antsiranana
Madagascar

Edoh TOSSOU
Laboratory of Mathematical
Modelling and Applications
Department of Mathematics
Faculty of Sciences
University of Lomé
Togo

Stéphane VINCENT
Laboratoire Modélisation et
Simulation Multi Echelle (MSME)
Université Gustave Eiffel
Paris
France

Roger VONY
Laboratoire de l'Energie
Renouvelable et Environnement
l'E.S. Polytechnique
Université d'Antsiranana
Madagascar